A Volume in The Laboratory Animal Pocket Reference Series

The Laboratory FERRET

The Laboratory Animal Pocket Reference Series

Series Editor

Mark A. Suckow, D.V.M.
Freimann Life Science Center
University of Notre Dame
Notre Dame, Indiana

Published Titles

The Laboratory Canine

The Laboratory Cat

The Laboratory Ferret

The Laboratory Guinea Pig, Second Edition

The Laboratory Hamster and Gerbil

The Laboratory Mouse

The Laboratory Nonhuman Primate

The Laboratory Rabbit, Second Edition

The Laboratory Rat

The Laboratory Small Ruminant

The Laboratory Swine, Second Edition

The Laboratory Xenopus sp.

The Laboratory Zebrafish

A Volume in The Laboratory Animal Pocket Reference Series

The Laboratory FERRET

C. Andrew Matchett • Rena Marr • Felipe M. Berard
Andrew G. Cawthon • Sonya P. Swing
Lovelace Respiratory Research Institute in Albuquerque, NM

CRC Press
Taylor & Francis Group
Boca Raton London New York

CRC Press is an imprint of the
Taylor & Francis Group, an **informa** business

CRC Press
Taylor & Francis Group
6000 Broken Sound Parkway NW, Suite 300
Boca Raton, FL 33487-2742

Printed in the United States of America on acid-free paper
Version Date: 20120124

International Standard Book Number: 978-1-4398-6181-3 (Paperback)

Visit the Taylor & Francis Web site at
http://www.taylorandfrancis.com

and the CRC Press Web site at
http://www.crcpress.com

Dedicated to the memory of Patricia J. Marx. Her encouragement, grace, and presence are greatly missed.

Contents

acknowledgments

The authors would like to acknowledge Steven Allen and Krystle Pacheco for their time and stalwart support. It is their effort that has produced many of the fine images included in this volume. Andrew Gigliotti, DVM, PhD, DCVP, graciously provided invaluable pathology data. We would also like to thank our coworkers, friends, and families. Without their patience and support, this work would not have been possible.

the authors

C. Andrew Matchett, DVM, is a clinical veterinarian at the Lovelace Respiratory Research Institute in Albuquerque, New Mexico. Dr. Matchett graduated from the veterinary school at Tufts University in 2002. He then completed a postdoctoral fellowship in Laboratory Animal Medicine at the Emory University School of Medicine and Yerkes National Primate Research Center in Atlanta, Georgia, in 2004.

Dr. Matchett's professional interests and background include nonhuman primate medicine and surgery, minimally invasive surgery and diagnostic imaging of exotic species, and respiratory disease models. He has memberships in the Association of Primate Veterinarians and AALAS.

Rena D. Marr, RLATG, is the husbandry services manager at Lovelace Respiratory Research Institute, Albuquerque, New Mexico. Ms. Marr received her undergraduate degree in agriculture from New Mexico State University. She is a certified laboratory animal technologist and is an active member of the IACUC. Her daily responsibilities include managing the animal husbandry staff and animal facilities.

Felipe Berard, DVM, received his Doctor in Veterinary Medicine degree from National University of Buenos Aires, Argentina, in 1983. He completed its laboratory animal internship in 1985. In 2000 he became a diplomat on veterinary education at Catholic University of Salta, Argentina. Since 1985 he has held clinical veterinarian positions in industry and in academia in Argentina and the United States. Dr. Berard is currently a clinical veterinarian at Lovelace Respiratory Research Institute (LRRI). He is responsible for the veterinary care of many species of animals and has other duties, including scientific assistance in research studies, training, education, and serves as a member of IACUC. He is a member of the American Association for Laboratory Animal Practitioners (ASLAP).

Andrew Cawthon, PhD, received his doctorate in microbiology and immunology from Wake Forest University School of Medicine in 2002. He entered a postdoctoral fellowship at Columbus Children's Research Institute (now Nationwide Children's Hospital) in Columbus, Ohio, studying hepatitis C virus infection in chimpanzees and humans. In 2006 he became a principal research scientist at Battelle Biomedical Research Center in Columbus, Ohio, assisting in the development of a ferret animal model for the evaluation of vaccines and therapeutics against influenza. In 2008, he joined Lovelace Respiratory Research Institute, serving as the director of preclinical microbiology and immunotoxicology, where he further developed the ferret model, applying that research to further evaluate anti-influenza therapeutics, while also developing training programs and curricula for working with ferrets in a research setting.

Sonya P. Swing, DVM, PhD, DACLAM, received her veterinary degree from North Carolina State University and earned a doctorate from the University of Alabama at Birmingham in experimental pathology. Dr. Swing is board certified by the American College of Laboratory Animal Medicine. She has managed departments of laboratory animal care and medicine in academia and in business. She is currently consulting through Camelot Vet Consulting Services.

1

important biological features

introduction

From the cargo hold of 18th-century naval vessels, to labyrinthine rabbit burrows, to cutting-edge influenza research, few animals have provided as much unheralded contribution to the state of human health as the domestic ferret (**Figure 1**). The ferret's prowess as a hunter has been well established for centuries. The ability of this tenacious mustelid to successfully seek out elusive quarry (most notably rabbit and rodent species) under challenging conditions prompted the Royal Navy to christen 15 warships *H.M.S. Ferret* between 1704 and 1940.[1] While ferrets were prized and actively used to mitigate the rodent threat to shipboard food supply, the scientific knowledge of the day would not have permitted a full appreciation of the public health implications of controlling species now known as key reservoirs of zoonotic disease.

It is fitting that an animal renowned for its inquisitive nature is helping advance the current state of medical knowledge. The biological characteristics of this unique animal have allowed it to emerge as an important laboratory animal model for the study of human viral respiratory diseases such as influenza. H1N1 (swine flu) and H5N1 (highly pathogenic avian influenza) are diseases of global public health interest that are currently being studied in the ferret with an emphasis toward the development of vaccines and therapeutics[2,3] (**Figure 2**). Ferrets are also employed to research areas as diverse as reproductive biology, pediatric intubation training, cystic fibrosis, and *Helicobacter* studies.[4,5,6] From pests to pestilence, the creature has captured the imagination of researchers seeking to "ferret out" the underpinnings of disease and develop new treatments

Fig. 1 Pair-housed ferrets.

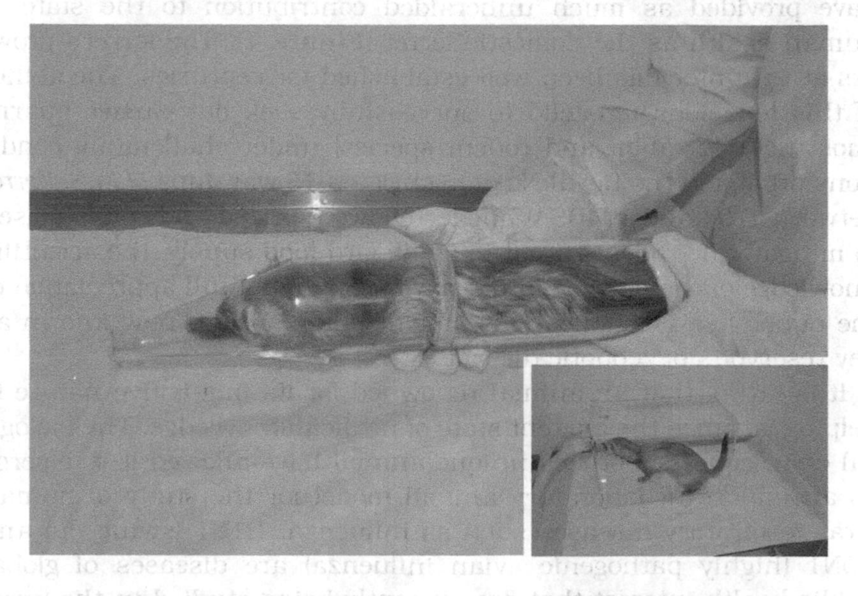

Fig. 2 Ferret in restraint tube and induction chamber (inset).

and interventions that may benefit human and veterinary medicine alike. Not bad for an animal that has been domesticated for over 2,000 years, yet is omitted from the U.S. Animal Welfare Regulations and glaringly absent from numerous mainstream sources of guidance for laboratory animal use.

taxonomy

The domestic ferret (*Mustela putorius furo*) is a carnivore belonging to the family Mustelidae (other common members include weasel, mink, marten, otter, and skunk). It is believed to have been derived from either the European polecat (*Mustela putorius*) or the steppe polecat (*Mustela eversmanni*). The carnivorous mammals belonging to this family share a number of common characteristics, including elongated bodies, short legs, thick coats, anal scent glands, delayed embryonic implantation (although this is not seen in the domestic ferret), and developed carnassial and sectorial teeth (the premolars and the first lower molar, used to shear flesh).

behavior

- Ferrets are often characterized as playful animals. They are generally not aggressive and relatively easy to handle. Caution should be exercised, however, when working with these animals in a research environment, as experimental factors may alter their ordinary responses to stimuli such as handling (**Figure 3**).
- Ferrets are neither strictly diurnal nor nocturnal. While there is evidence suggesting ferrets do have circadian cycles influenced by photoperiod (length of light and dark cycles), it appears not to affect their pattern of activity. Pet ferret owners may attest to this anecdotally, as ferrets often adapt their activity levels in accordance with those of their human housemates.
- Ferrets prefer to defecate in a corner. This is usually preceded by the raising of the tail and a backwards shuffling motion.
- Ferrets are seasonal breeders, with the onset of estrus commencing during longer light cycles in the photoperiod.
- Ferrets are known to hoard food and objects of interest.

key anatomic and physiological features

For an excellent, comprehensive description of ferret anatomy, see Fox's *Biology and Diseases of the Ferret*.[7] Many of the important and

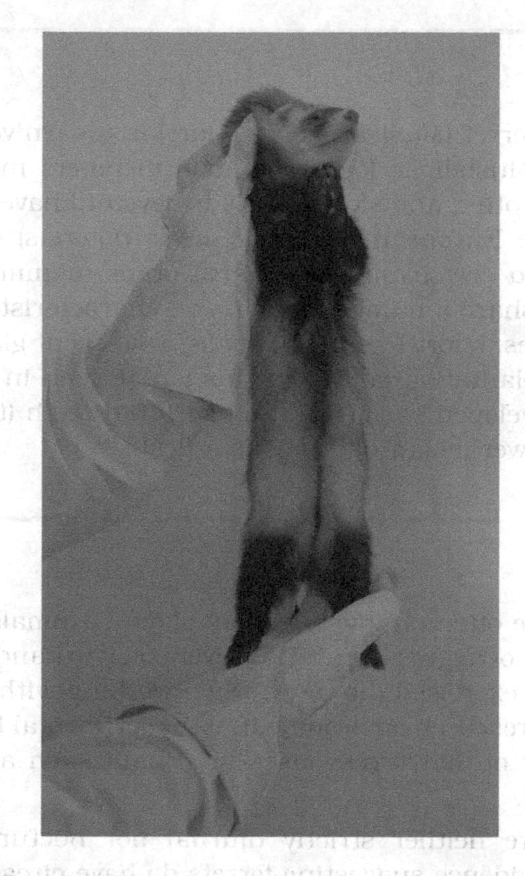

Fig. 3 "Scruffing" a ferret.

clinically salient anatomic and physiological features that distinguish them from other small laboratory animals are detailed below.

General Characteristics

- Similar to other Mustelidae, ferrets have a thick coat, short legs, and a narrow, elongated body well suited to burrowing.
- Ferrets possess anal scent glands and sebaceous glands, but lack developed sweat glands. Although less pronounced than that of other mustelids, ferrets will retain a musky odor post-neutering and de-scenting from their sebaceous secretion. The absence of developed sweat glands makes them particularly susceptible to heat.
- Ferrets cope best at ambient temperatures of 15–25°C (59–77°F), but are subject to heat prostration at temperatures above 30°C (85°F).

- Male ferrets (hobs) are larger than females (jills), with weights typically ranging up to 2 kg and approximately 1 kg, respectively. Seasonal weight fluctuations do occur. The young (kits) are an average of 8–10 g at birth (whelping).
- A typical ferret lifespan is 6–8 years, although some individuals may live several years longer.

Coat Color Variation

Numerous coat color variations exist in commercially available ferrets, some of which have clinical implications for the researcher.

- Sable or Fitch—characterized by a yellowish coat with dark patches over the extremities (limbs and tail), a well-defined facial mask, and dark eyes (typically brown).
- Silver Mitt—similar to a Sable, but with white markings on the feet (hence "Mitt") and chest. The mask is typically less pronounced than that of the Sable. Breeding Silver Mitts with one another may result in congenital defects and increased fetal loss.
- Siamese—have brown guard hair. "Cinnamon" and "Chocolate" are often used to describe Siamese coats with red-brown and dark-brown markings, respectively.
- Albino—have a white coat and red eyes.
- Black-eyed White—have a white coat and dark eyes. Cross-mating Black-eyed Whites may produce congenital defects and increased fetal loss.

There is a coat color association in ferrets with the occurrence of Waardenburg syndrome. This syndrome (resulting from the autosomal dominant inheritance of one of a number of altered genes) is often associated with varying degrees of deafness and partial albinism or other pigmentary alterations in both people and ferrets. A large percentage of the ferrets with the following patterns are deaf; care should be exercised to avoid startling them when handling is necessary. These patterns may be superimposed on ferrets with the aforementioned coat color variations:

- Panda—a white head (lacking a mask) and a white bib.
- Blaze—a white stripe on top of the head (similar to that on a badger) and a white bib.

Skeletal

- Ferrets have 7 cervical, 15 thoracic, 5 to 7 lumbar, 3 sacral, and 18 caudal vertebrae. There are commonly 15 pairs of ribs, with the cranial 10 articulating distally with the sternum and the caudal 5 pairs terminating in the costal margin. The thorax widens as it extends caudally. This conformation allows for a compliant thoracic wall.
- The forelimbs and hindlimbs have 5 toes each. All digits have a non-retractable claw distally.

Dentition

- The dental formula of the adult ferret is 2 (3/3 incisors, 1/1 canines, 4/3 premolars, 1/2 molars).
- The deciduous teeth erupt during the 3rd to 4th week of life.
- The permanent teeth are all erupted before the end of the 3rd month of life.
- Like all mustelids, ferrets lack a second upper molar.

Respiratory

- Ferrets possess a long trachea with incomplete cartilaginous rings. The laryngeal anatomy renders endotracheal intubation difficult, making the animal an excellent subject on which to practice neonatal and infant intubation.
- The left lung consists of a cranial and a caudal lobe; the right lung has a cranial, a middle, a caudal, and an accessory lobe. The lungs extend approximately from the 2nd intercostal space to the apices bilaterally around the 11th intercostal space.
- The combination of a relatively longer, wider diameter trachea and compliant thoracic anatomy allows for decreased airway resistance and a relatively large lung capacity. These factors are ascribed as burrowing adaptations that are of potential interest to respiratory researchers.[7]

Note: Ferrets are susceptible to numerous human respiratory viruses (including influenza A and B) and often exhibit clinical signs similar to those seen in humans. Consequently, ferrets may not only acquire respiratory viruses from animal caretakers, but transmit viruses to susceptible workers as well. N95 respirators (without one-way relief valves) are often used not only to safeguard the health of laboratory animal workers, but also to prevent the inadvertent exposure of study ferrets to human respiratory viruses that may confound viral respiratory research (**Figure 4**).

- The emergence of the ferret as the model of choice for influenza research and concerns about cross-contamination with unrelated respiratory pathogens have lead to the development of specific-pathogen-free (influenza) barrier ferret production facilities. Animals from these facilities are often not well socialized to people, and care should be taken while handling such ferrets.

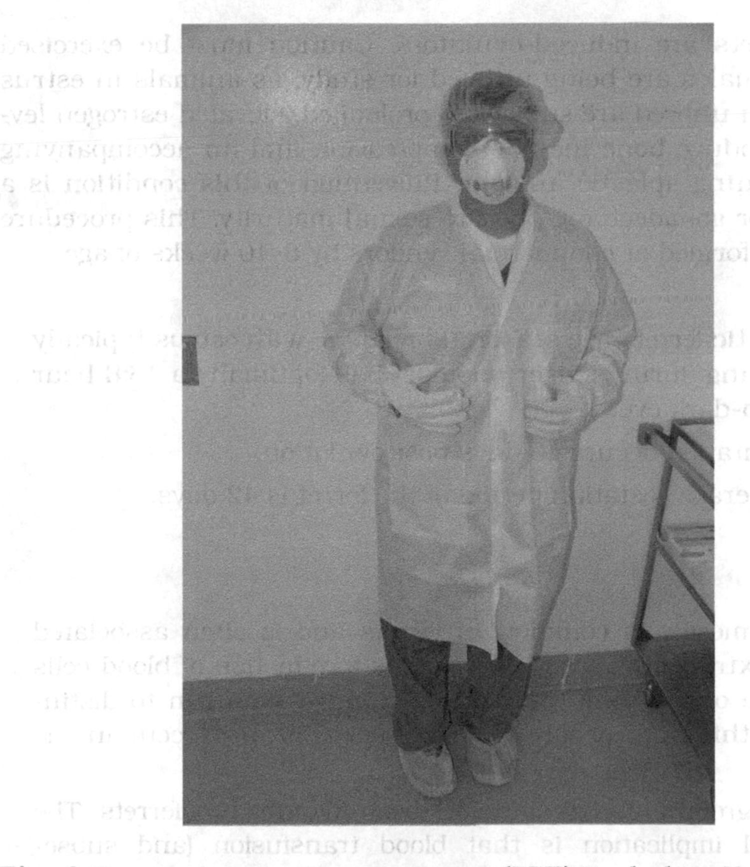

Fig. 4 Personal protective equipment (PPE), including N-95 respirator (without relief valve) and bite-resistant gloves.

Gastrointestinal

- The ferret is an obligate carnivore. Dietary requirements include high fat and protein intake and low fiber levels.
- The ferret is monogastric, and lacks a cecum; the ileocecal junction is not discernable on gross examination.
- Gastrointestinal transit time is short (typically less than 3 hours).

Urogenital and Reproductive

- Females have a bicornuate uterus.
- Males have an os penis, two testes, and an ill-defined prostate.
- Ferrets reach sexual maturity between 4 and 8 months of age.

Note: Ferrets are induced-ovulators. Caution must be exercised if intact females are being utilized for study, as animals in estrus that remain unbred are subject to prolonged, elevated estrogen levels that produce bone marrow suppression and an accompanying life-threatening aplastic anemia. Prevention of this condition is a rationale for gonadectomy prior to sexual maturity. This procedure is often performed at commercial vendors by 8–10 weeks of age.

- Domestic ferrets are seasonal breeders, with estrus typically occurring during longer photoperiods (optimally a 16:8 hour light-to-dark cycle).
- Implantation occurs 13 days post-ovulation.
- The average gestation period of the ferret is 42 days.

Hemolymphatic

- Splenomegaly is common in ferrets and is often associated with extramedullary hematopoiesis (production of blood cells outside of the bone marrow). Care must be taken to distinguish this from neoplastic processes that also occur in the ferret.
- Blood group antigens have not been identified in ferrets. The clinical implication is that blood transfusion (and subsequent repeat transfusion) may be safely performed without crossmatching.[8]

Normative Values

A summary of normal biological parameters of the ferret can be found in **Table 1**.

Clinical Pathology

Marshall Farms has long been the principal, well-respected commercial vendor of ferrets for research. For more in-depth information on normal hematology and clinical chemistry values in ferrets from that source (delineated by age, gender, and coat color), please consult *Biology and Diseases of the Ferret*[7] or *The Marshall Bioresources Reference Data Guide*.[9]

As new focus on ferret use in the field of respiratory research has recently developed, the authors have included normative data from male ferrets from an influenza specific-pathogen-free barrier facility (male ferrets are currently favored for such research, as their larger body mass allows for increased blood sampling and easier handling). Marshall Farms' values are cited for purposes of comparison. The data suggest a fair amount of consistency in mean values between the commonly used populations.

Hematology

The data presented here were compiled at the Lovelace Respiratory Research Institute (LRRI, Albuquerque, NM) from 279 clinically normal Triple-F Farms (Sayre, PA)–sourced ferrets (**Table 2**). Values are relatively consistent across both populations of males (predominantly in the 4–6 month age range). Some minor differences in red blood cell count and morphology may be attributable to the ferrets acclimating to the altitude at LRRI of approximately 1600 m (5300 ft) above sea level.

TABLE 1: NORMATIVE BIOLOGIC DATA FOR FERRETS[7,10]

Parameter	Representative Value	Reference(s)
Lifespan (years)	6–8	Fox 1998, Johnson-Delaney 1996
Weight (kg)	♂ ≤ 2 kg, ♀ ≤ 1 kg	Fox 1998, Johnson-Delaney 1996
Heart rate (beats/min)	200–400	Fox 1998
Respiratory rate (breaths/min)	33–36	Fox 1998
Temperature (°C)	37.8–40.0	Fox 1998
Food intake (g/day)	20–40	Johnson-Delaney 1996
Water intake (mL/day)	75–100	Fox 1998
Urine production (mL/day)	26–28	Fox 1998

Table 2: Hematology Data in Male Ferrets

Parameter	WBC (x10E3 cells/µL)	RBC (x10E6 cells/µL)	HGB (g/dL)	HCT (%)	MCV (fL)	MCH (pg)	MCHC (g/dL)	RDW (%)	HDW (g/dL)
					Marshall Farms				
Low	7.8	6.9	11.4	38.1	49.8	15.2	28.0	11.9	1.4
High	17.2	9.4	15.2	51.1	58.9	17.2	31.6	14.4	1.6
Mean	12.6	8.1	13.3	45.2	54.9	16.2	29.5	13.2	1.5
SD	2.5	0.7	0.9	2.8	1.8	0.6	0.4	0.7	0.1
					Triple-F Farms				
Low	4.81	5.9	10.8	34.6	50.3	16.4	29.2	11.0	1.3
High	15.86	10.8	18.2	57.2	62.7	19.5	34.1	14.6	1.9
Mean	9.37	7.7	13.7	43.6	56.6	17.9	31.6	12.6	1.6
SD	2.98	1.2	1.7	5.6	3.1	0.9	1.3	0.9	0.1

Parameter	NEUT (%)	LYMPH (%)	MONO (%)	EOS (%)	BASO (%)	NEUT-ABS (x10E03 cells/µL)	LYMPH-ABS (x10E03 cells/µL)	MONO-ABS (x10E03 cells/µL)	EOS-ABS (x10E03 cells/µL)	BASO-ABS (x10E03 cells/µL)	PLT (x10E03 cells/µL)	MPV (fL)
						Marshall Farms						
Low	19.5	37.6	0.6	2.0	0.2	2.2	4.0	0.1	0.2	0.0	430.9	4.9
High	55.0	74.5	2.4	6.7	0.7	8.1	10.2	0.3	0.7	0.1	1083.2	15.1
Mean	37.3	56.2	1.4	4.0	0.3	4.8	7.0	0.2	0.5	0.0	783.3	8.4
SD	10.6	10.7	0.6	1.6	0.1	2.1	1.7	0.1	0.1	0.0	130.6	0.9
						Triple-F Farms						
Low	16.4	36.0	1.5	1.0	0.2	1.3	2.4	0.1	0.1	0.0	254.0	6.7
High	54.2	77.1	10.4	5.5	0.9	6.6	10.5	0.9	0.5	0.1	888.0	13.6
Mean	32.4	60.1	3.4	2.6	0.5	3.0	5.6	0.3	0.2	0.0	585.4	8.9
SD	9.5	10.9	2.1	1.1	0.2	1.3	2.0	0.2	0.1	0.0	150.5	1.6

TABLE 3: SERUM CHEMISTRY DATA IN MALE FERRETS

Parameter	NA (mmol/L)	K (mmol/L)	CL (mmol/L)	GLU (mg/dL)	BUN (mg/dL)	CREAT (mg/dL)	PHOS (mg/dL)	BILI-T (mg/dL)	ALP (IU/L)
					Marshall Farms				
Low	143.0	4.9	111.8	91.6	24.4	0.5	7.3	0.1	66.2
High	157.0	5.8	123.0	135.2	42.3	1.0	10.3	0.5	153.3
Mean	147.0	5.5	116.4	118.1	32.0	0.8	9.0	0.3	109.8
SD	2.4	0.2	2.2	10.0	4.1	0.1	0.7	0.1	24.4
					Triple-F Farms				
Low	141.0	3.9	106.2	81.2	14.0	0.1	4.7	0.0	16.1
High	151.0	5.1	119.8	177.8	39.8	0.7	12.5	0.1	211.5
Mean	146.0	4.5	113.3	119.0	26.4	0.4	8.8	0.1	115.7
SD	2.7	0.3	3.1	22.8	7.0	0.2	2.0	0.0	55.4

Parameter	ALT (IU/L)	AST (IU/L)	GGT (IU/L)	TP (g/dL)	ALB (g/dL)	CA (mg/dL)	CK (IU/L)	CHOL (mg/dL)	GLOBN (g/dL)
					Marshall Farms				
Low	93.0	61.4	5.0	5.3	2.6	9.7	138.0	137.1	2.4
High	381.6	122.2	17.6	6.7	3.3	10.8	454.5	227.9	3.7
Mean	243.8	94.6	9.1	6.1	3.0	10.2	307.5	174.8	3.1
SD	85.4	13.9	3.4	0.3	0.2	0.3	90.6	24.1	0.3
					Triple-F Farms				
Low	79.9	54.1	0.0	4.7	2.8	8.8	157.1	90.2	1.8
High	566.5	188.0	12.9	6.5	3.8	10.4	1408.8	265.7	3.0
Mean	218.8	98.0	3.7	5.6	3.2	9.6	600.3	185.8	2.3
SD	119.9	31.6	3.4	0.4	0.3	0.4	302.1	47.5	0.5

TABLE 4: CEREBROSPINAL FLUID VALUES IN ADULT FERRETS[11]

Component	Range, Mean Value	Reference
White blood cells (cells/μL)	0–8, 1.6	Platt et al. 2004
Red blood cells (cells/μL)	20.3	Platt et al. 2004
Protein (mg/dL)	28.0–68.0, 31.4	Platt et al. 2004

Clinical Chemistry

The serum chemistry data presented here was compiled at LRRI from 160 clinically normal Triple-F Farms–sourced ferrets (**Table 3**). Marshall Farms' data are included for purposes of comparison.

Miscellaneous Clinical Parameters

Cerebrospinal fluid values for adult ferrets can be found in **Table 4**.

2

husbandry

introduction

Healthy animals and consequently good husbandry practices are essential for the quality of the biomedical research being conducted. Effective husbandry practices will address both the macroenvironment (room and surroundings) and the microenvironment (immediate caging or housing area) as well as nutrition/feeding, social, and enrichment requirements of the animals.

housing

Currently, there are no standards described in the National Research Council's *Guide for the Care and Use of Laboratory Animals* (the *Guide*) nor in the USDA Animal Welfare Regulations for ferret caging requirements, and questions about ferret caging are common. Caging, in general, needs to provide sufficient space to allow animals to make postural and social adjustments with freedom of movement. Caging materials for ferrets can be almost anything that is easily sanitized and can withstand the abuse the animals will mete out.[12] Several references indicate that two adult animals can occupy a home cage of 24 x 24 x 18 inches (60 x 60 x 46 centimeters).[13,14] Stainless steel rabbit caging as well as the molded-bucket-style rabbit caging have been successfully modified to house ferrets. Particular attention needs to be paid to the width of both the cage bars and the floor spacing. Ferrets can get through small spaces, so the bar spacing should be

1 in. (2.5 cm) for adults to thwart potential escapees[7,13,15] (**Figure 5**). Bar spacing will need to be adjusted for young and smaller ferrets. Ferrets generally use specific corners to defecate, so the floor grids need to be wide enough to allow excreta to fall through but small enough to be comfortable on the animals' feet. Providing a nesting box with some type of bedding can help relieve stress in animals and will allow them a place to get away from the bustle of the room. Bedding can be old newspaper or non-aromatic wood chips. Pine and cedar wood chips are not recommended because the oils can cause irritation to the animals' eyes and respiratory system. Some personnel have used large polycarbonate rat breeding cages with bedding for short-term housing and for transporting ferrets. This type of caging should be used only for short-term housing, and the caging should be dedicated to ferrets because odor will be retained in the caging.[13] Solid-wall caging, such as rodent cages and aquariums or fish tanks, should not be used for home cages for ferrets because they do not allow adequate ventilation.[14]

environmental conditions

Ferrets are sensitive to heat and should be housed in temperatures between 4° and 25°C (40° to 77°F).[7] They are susceptible to heat stress at elevated temperatures, and temperatures approaching 27°C (80°F)

Fig. 5 Bar spacing on caging.

can produce heat stroke. Temperatures of 32°C (90°F) can be fatal.[16] Adult animals are more susceptible to higher heat than younger animals owing to their poorly developed sweat glands and thin subcutaneous fat layer.[7] If using tiered caging systems or racks, animals on the upper levels should be monitored closely for heat stress. Humidity does not seem to be a significant environmental factor except in the presence of chronically elevated temperatures.[13,14,17]

Room ventilation should be 10–15 complete non-recirculated air changes per hour. Air changes at the higher end of this range are beneficial because of the animals' musky odor and susceptibility to respiratory diseases.[7,13] Care should be taken to keep animals out of drafts and continuously blowing hot or cold air, as this can cause ferrets to become sick.[17]

Room illumination providing 12 hours of light and 12 hours of dark are sufficient to maintain animals in conventional housing.[7] It is important to have cages or rooms that can be completely darkened at night. Incomplete dark cycles or cycles that are interrupted with light can create health problems in ferrets.[17]

Noise levels in the ferret housing area should be kept at a minimum. Loud and sudden noises can stress the animals and interrupt sleep cycles.[17] As such, it is suggested that ferret housing areas be kept away from mechanical areas such as cage wash areas.

environmental enrichment

The primary focus of environmental enrichment is to provide mental and physical stimulation that promotes the animals' well-being.[18] Social housing is by far the most effective form of environmental enrichment for the ferret (**Figure 6**). Animals can be housed in same-sex groups, and even intact males can be group housed with a little care. Space requirements are dependent on the number of animals being housed together. Space should be sufficient to allow for normal movements and postural adjustments. In group housing situations, animals will play with each other and may even drag each other around by the nape of the neck. This is not cause for concern unless the activities get over zealous.[7,15,18] When animals need to be single housed, they should be housed in a way that allows olfactory and auditory stimulation between individuals.[13]

Toys also provide enrichment for the ferrets, but care should be taken concerning what types of toys are offered. Toys made of soft rubber or pieces that can be easily chewed off should not be offered to

Fig. 6 Social-housed ferrets.

ferrets as these can be ingested and cause intestinal blockage. Suitable toys include items made of hard rubber or plastic, polyvinyl chloride (PVC) piping, and clothes dryer hosing. Even items such as large mailing tubes and paper bags make excellent ferret toys. Any toys given to ferrets should be checked frequently for use and wear[14,17] (**Figure 7**).

Ferrets are strict carnivores, and as such they cannot use carbohydrates effectively or digest fiber. These points need to be taken into account when deciding what a ferret will be offered as a food treat.[14] Some food items that can be offered to ferrets are eggs, meat, insects, small prey animals, and chicken baby food. Enrichment food items should always be offered sparingly so the digestive tract of the animal is not disturbed.

nutritional requirements

As mentioned earlier, ferrets are strict carnivores and are designed to eat frequent and small meals of meat proteins. They cannot digest fiber or carbohydrates effectively.[14] An adult animal will eat about 43 g/kg body weight of dry food a day.[7] The ferret will eat up to 10 meals a day if given feed ad libitum. Care should be taken that animals do not become obese if offered food all the time. The nutritional requirements lean heavily toward crude, meat-based protein making up 30–40% of the diet.[7,14,15] The diet should also include approximately 18–30% fat.[3,7,14]

Fig. 7 Ferrets with toys/environmental enrichment in cage.

There are several commercially available ferret diets on the market that meet the unique dietary requirements for the ferret. An alternative to the commercially available diets is to feed the ferret a high-quality dry cat food supplemented with canned varieties of high-quality cat food.[7] Most ferrets seem to enjoy eating various fruits, but these foods should be given in limited quantities and with caution as the animal will over eat the fruit in lieu of a healthier diet if given the choice.[14]

Clean fresh water should be offered ad libitum. Water can be supplied in heavy crocks, stainless steel bowls, water bottles, or via automatic watering systems. The water bottle and the automatic waterer tend to be the preferred methods because they preclude the animal playing with and dumping the water out of a bowl. Ferrets also tend to adapt to the automatic watering systems and water bottles relatively quickly and well. Regardless of the method given, water should be changed daily or the lixit systems on automatic watering systems checked daily.[14]

sanitation

Caging should be cleaned out at least daily and sanitized with appropriate methods weekly to keep animals healthy. Daily cleaning should include removing excreta. Ferrets can usually be trained to use a litter box quite easily, and this facilitates daily cleaning of the cage. If using litter boxes, the wall height should accommodate the ferret's tendency to back into the corner to defecate. There are several litters available for use, but it is recommended that clay litter not be used, to avoid getting wet clay on the animal. Litter boxes, if used, should be changed out daily with fresh litter.[7,13,14]

Various methods of sanitation exist, but all should kill microorganisms. Cages can be sent through a commercial cage wash unit that uses both chemicals and heat to kill vegetative organisms. After strict chemical sanitation, caging should be rinsed thoroughly and dried before reintroducing animals. A variety of chemicals are available for daily cleaning as well as sanitation, and the product used should be selected based on effectiveness and experience.

The room as well as the caging needs to be cleaned. Daily sweeping and mopping will help keep debris and fur from gathering on floors and in corners. Supply and exhaust vents should also be cleaned regularly. The entire room should be sanitized on a regular schedule to prevent the possible spread of microorganisms.

The sanitation methods and practices in use should be periodically assessed. This can be performed by several methods. Temperature tapes indicate by a change in color whether thermal-disinfection equipment is working properly and so will ensure that caging is getting properly sanitized. Monitoring the cage-wash areas using temperature tapes is usually done daily. Microbial monitoring can be performed by using a RODAC (Replicate Organism Detection and Counting) plate on equipment and rooms that have been recently sanitized. The RODAC plate is placed on the area to be checked and incubated for 48 hours; the colonies formed are then counted. Bacteria growth in the moderate to heavy range can detect a weakness in the sanitation process or a possible equipment failure. A third method of checking sanitation procedures is to use an adenosine triphosphate (ATP) bioluminescence reader. This method consists of swabbing an area to be checked and inserting the swab into the reader, which shows results almost immediately. Facilities are responsible for determining their own levels of acceptable ATP levels. This method of quality control provides a quick measure of how sanitation programs and methods are working.

transportation

Ferrets can be delivered to facilities by ground or air, or a combination of both. Ideally, door-to-door transportation from the vendor to the research facility occurs. There are no specific regulations on container

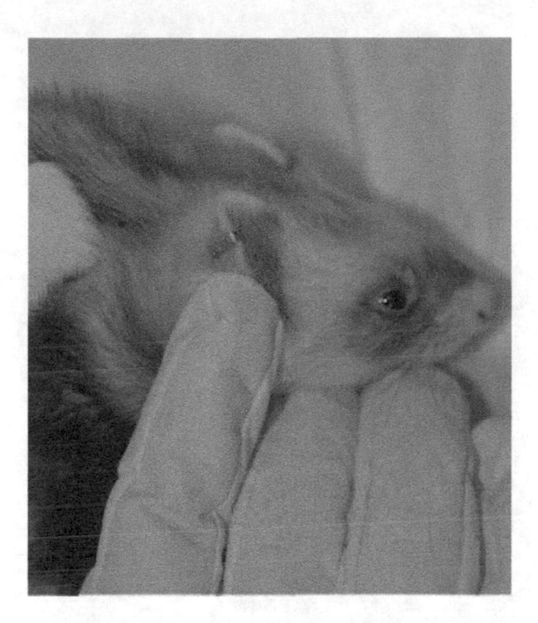

Fig. 8 Ferret with ear tag.

types or stocking densities for transporting ferrets, but animals must be able to lie down and make postural adjustments in the container. Containers for conventionally reared animals can include absorbent contact bedding for the trip and should be supplied with a feed and water or hydrogel source. Adequate ventilation must be available on the containers. Animals should be visible by caretakers, and containers should be escape proof. Care must be taken when shipping ferrets in warmer weather as animals can easily overheat during transit.[19]

record keeping

The accuracy and accessibility of animal records are essential for good husbandry practices. Ferrets can be identified by cage card, ear tattoo, or microchip, and this information should be recorded in individual animal records (**Figure 8**). Room and husbandry records should be maintained and should contain health checks, room and cage sanitation or changes, animal census in the room, and initials/date of the person performing the work. Room environmental records should also be maintained and kept. Individual health records should be documented as needed for medical treatments given, body weights obtained, medical procedures performed, etc.[20] In all cases, records should be readily available and understood when referenced.

Management

regulatory and compliance

Regulatory and compliance issues discussed below are pertinent within the United States.

United States Department of Agriculture (USDA)

The laboratory ferret is not specifically addressed within the Animal Welfare Act and Animal Welfare Regulations[12] administered by the USDA. Thus the ferret is regulated under Subpart F— Specifications for the Humane Handling, Care, Treatment, and Transportation of Warmblooded Animals Other Than Dogs, Cats, Rabbits, Hamsters, Guinea Pigs, Nonhuman Primates, and Marine Mammals. This section of the regulations provides generic information on facilities, space requirements, feed, water, sanitation, employees, and separation. No specifics regarding temperature, cage size, or feed composition and amount are provided. Compatibility is the primary factor regarding cage density. The majority of information provided relates to transportation standards, including primary enclosures, care, handling, food and water during transit, intermediate carriers, and methods of transportation.

Public Health Service (PHS)

The laboratory ferret falls within the definition of *animal* as used by the PHS Policy on Humane Care and Use of Laboratory Animals, wherein "animal" is "any live, vertebrate animal used or intended for use in research, research training, experimentation or biological testing or for related purposes."[21] Any entity that receives PHS funding to perform ferret research will need to have an Animal Welfare Assurance Statement on file with the National Institutes of Health (NIH) Office of Laboratory Animal Welfare (OLAW). The Health Research Extension Act of 1985[22] is the legislation that requires guidelines to be set for the proper care and treatment of animals used in research, including the use of anesthesia, analgesia, tranquilizers, paralytics, and euthanasia and the pre- and post-surgical monitoring of animals undergoing research surgery. The PHS Policy mandates the existence of an Institutional Animal Care and Use Committee (IACUC) and its composition, duties, and review of protocols describing the research use of animals.

Food and Drug Administration (FDA) and Environmental Protection Agency (EPA) Good Laboratory Practice (GLP)

GLP standards regulated by the FDA in CFR Title 21 Part 59[23] specify requirements for research involving food additives, drugs, biological products, and medical devices intended for use in animals or humans. The ferret has been used extensively in GLP testing and has been used as a second species in data submitted to the FDA for approval.[13] One example of applicability for ferret research would be in the use of ferrets to develop influenza or SARS-CoV vaccines.[24] The GLP regulations do not address ferrets and their use in research specifically, but instead address the manner in which GLP studies must be conducted to assure consistency, reliability, and reproducibility of the results.

GLP standards established by the EPA govern research to assess the safety of chemicals on humans, animals, and the environment.[25] The domestic ferret commonly used in biomedical research has been used to establish the toxicity of pesticides and rodenticides to the endangered black-footed ferret.[26,27]

Association for Assessment and Accreditation of Laboratory Care International (AAALAC)

AAALAC is a voluntary, non-profit organization that accredits programs of animal care. Accreditation by AAALAC impacts the category of Institutional Status recognized by OLAW. As of autumn 2011, AAALAC uses three reference documents to evaluate animal programs during site visits: (1) *Guide for the Care and Use of Laboratory Animals*, 8th edition (the *Guide*),[18] (2) *Guide for the Care and Use of Agricultural Animals in Research and Teaching* (the Ag *Guide*),[28] and (3) the European Convention for Experimental and Other Scientific Purposes (ETS 123).[29] As the ferret is not considered an agricultural animal, the Ag Guide does not regulate ferret use in research and teaching.

The Guide

While the *Guide* does not specifically list the ferret, its overall program of animal care and many of the components of specific programs will be applicable to any institute performing research, testing, or teaching with the laboratory ferret. An accredited program must follow the recommendations of the IACUC and related organizations, including occupational health and safety, environmental management (of temperature, humidity, lighting, housing, etc.), veterinary care, and physical plant component considerations.

ETS 123

ETS 123 does have a section specific to the ferret: Appendix A, Species-Specific Section, E, Species-specific provisions for ferrets. This section defines ferret-specific requirements for temperature range, lighting, noise exposure, social housing, types of enrichment, and caging, including size, flooring, bedding, and sanitation. More general information is provided for ventilation, humidity, health, euthanasia, identification and record-keeping. ETS 123 notes that ferrets become nervous in the absence of auditory stimulation, but also develop stress disorders when exposed to loud, unfamiliar noises. It is also noted that ferrets benefit from social housing with the exception of intact males during breeding season and pregnant females within two weeks of parturition. Single housing must be only with scientific or welfare justification.

Institutional Animal Care and Use Committee (IACUC)

There are several factors the IACUC will need to evaluate when reviewing research protocols involving ferrets. The first involves the level of pain and distress associated with the described experiments. As ferrets are frequently used in studies that involve more than momentary pain or distress (infectious disease, toxicology), the IACUC must ensure that the appropriate pain/distress category is assigned and evaluate the use of anesthesia, analgesia, or tranquilization. Consultation with or review by a veterinarian is necessary for those protocols that involve significant pain or distress. A search for alternatives to the painful/distressful procedures must be performed, and a summary of the potential alternatives, if any, and why they cannot be used, must be included in the protocol. As described in the *Guide*, humane endpoints must be defined to minimize the pain or distress the ferret experiences. The researcher should establish these endpoints prior to performance of the experiment, ensure they are scientifically sound, and provide for the most humane endpoint. These endpoints are typically multifactorial, and may involve removal from the study, provision of pain/distress relief through medication, or euthanasia.

The second point that IACUC should pay particular attention to with ferret protocols is that of social housing. Ferrets benefit from social interaction[30] and should be housed in compatible social pairs or trios when possible. Strong scientific justification must be provided in the research protocol when single housing is requested. As stated previously, intact male ferrets and late gestation females should be housed singly as a matter of welfare.

A third item that should be noted with ferret protocols is any food restriction, including medical fasting, that is involved. While a wide range of fasting times have been published (see Chapter 4, "Veterinary Care"), the authors recommend a fasting period of 2–4 hours. Any procedures that require fasting should be carefully reviewed, as fasting for longer than 4 hours can lead to hypoglycemia in the ferret. Measuring blood glucose may be suggested in protocols that require prolonged dietary restriction.

State and Local Regulations

Each state has its own set of legislation regarding pet animals and animals used in research. Ferrets are addressed in the state's Department of Fish and Game Code. As of April 2011, ferrets are banned as pets in the states of California and Hawaii, and a permit must be obtained to import them for use in medical research.[31,32] Some states may not prohibit pet ownership but still require an import permit to bring ferrets into the state. Some cities such as New York and Washington, D.C., also ban pet ferret ownership. Each research, teaching, or testing facility should check with its state, county, and city to ensure compliance with these regulations.

Summary

With the exception of the ETS 123, regulations regarding the ferret for use as a laboratory animal pertain to the use of performance standards. These standards are intended to provide high-quality care of animals without the limitations of engineering standards. Within the ETS 123 there exist some engineering standards, such as a minimum cage size of 4500 cm^2. The research, teaching, or testing facility new to the use of ferrets will need to ensure that it maintains state and local compliance as well as federal and private (AAALAC) compliance.

zoonotic diseases

Viral Zoonoses

Two viral diseases are potential zoonotic threats to humans working with them, though one is in fact far more often an anthroponoses. Rabies is a potential zoonoses,[33] though it is highly unlikely, particularly in the laboratory-raised ferret that is vaccinated for rabies. The second viral disease is influenza types A and B.[34] These orthomyxoviridae are more likely to be communicated to the ferret from the human, but zoonotic potential also exists.

Bacterial Zoonoses

Campylobacteriosis is a potential threat for those working with ferrets. One report showed 18% of tested ferrets were positive for *Cambylobacter jejuni* and *C. coli*.[35] Campylobacteriosis is one of the most common bacterial infections in humans and causes severe gastroenteritis.

Ferrets may also carry *Salmonella* spp. (*S. typhimurium, S. hadar, S. kentucky*, and *S. enteritidis*), which can infect humans, resulting in diarrhea, fever, vomition, and abdominal cramping.[36] Salmonellosis is considered the most common zoonotic disease in humans, regardless of animal source.

Ferrets are very susceptible to mycobacterial infection, and *Mycobacterium avium, M. bovis*, and *M. tuberculosis* have all been isolated from ferrets. Cases have been reported primarily in Europe. Mycobacteriosis is highly unlikely in the laboratory ferret, however, due to modern management practices used in laboratory animal vivarian.

Parasitic Zoonoses

Cryptosporidium parvum, a non-host-specific protozoan, has been identified in the ferret. A 1988 report demonstrated that 31 of 44 ferrets at a biomedical research facility had cryptosporidiosis.[37] The infection is usually asymptomatic in ferrets, and people with an immunodeficiency or immunosuppression are at increased risk of infection and subsequent disease. *Giardia lamblia* (also known as *G. intestinalis*) is a second protozoan parasite that has been identified in ferrets[38] and can be communicated to humans.

Mycotic Zoonoses

Dermatophytosis (ringworm) is uncommon in the ferret, but infection with *Trichophyton mentagrophytes* has been reported.[39] Affected animals may be asymptomatic or have focal alopecia with or without pruritis.

occupational health

An occupational health program is a standard part of an accredited, inspected animal facility; however, a few items that are specific to the ferret should be noted. Since the ferret is a popular model for influenza and other respiratory viral studies and is extremely susceptible to infection by influenza, the occupational health program must address this. All personnel working with or having any direct exposure to ferrets should be vaccinated for influenza. Additionally, they should wear an N95 respirator without a one-way relief valve or other appropriate PPE to further minimize the risk of exposure to humans and the research ferret (**Figure 4**). Personnel with flu-like symptoms should report to their supervisor for evaluation prior to initiating contact with ferrets.

A second consideration for work with ferrets is their extreme susceptibility to tuberculosis (TB) and other mycobacterial infections. With global travel to TB-endemic areas, personnel should be tested and cleared prior to working with ferrets. Frequency of TB testing and the type of clearance (two-step negatives versus one negative) should be based on a risk assessment at your specific facility and should include other susceptible animal species, the potential for infection of personnel, and the frequency of additions to the animal colony.

Personnel in contact with ferrets may develop allergies. These may manifest as symptoms of conjunctivitis, allergic rhinitis, asthma, or skin rashes. Numerous proteins may cause ferret allergy, although albumin, a blood protein, is commonly thought to be the primary causal factor. Males produce more albumin in their urine and can elicit a stronger allergic reaction than females. Treatment for ferret allergies includes medications commonly used to treat cat allergy, and the most effective treatment is immunotherapy. Over-the-counter antihistamines also alleviate allergy symptoms. The personal protective equipment used when working with ferrets, specifically N95 respirators, may reduce exposure to ferret allergen.[40,41]

Zoonotic risk to personnel is minimized by proper use of personal protective equipment (**Figure 4**). Ferret bites or scratches can go through latex, vinyl, or nitrile gloves, and bite/scratch protectant gloves or sleeves are recommended for those performing manual restraint. Frequent, positive handling of the ferrets also reduces the risk of injury to both personnel and ferrets (**Figure 3**).

veterinary care

basic veterinary supplies

The following list is not comprehensive, being intended rather to provide the operator with a template from which to collect an appropriate armamentarium of supplies necessary for physical examination of ferrets. Additional materials may be required depending upon study needs (**Figures 9** and **10**).

- Stethoscope – a pediatric or neonatology model (e.g., Littmann Classic II Infant) is recommended
- Small-animal thermometer and disposable sleeves (plastic thermometers are preferred)
- Lubricating gel
- Disposable syringes (1–10 mL)
- Disposable hypodermic needles (20–23 G x 1")
- Blood collection tubes – serum separator/anticoagulant (e.g., EDTA for whole blood)
- Gauze sponges – 2 x 2's or 4 x 4's
- Light source – penlight or trans-illuminator
- Topical disinfectant – povidone-iodine, chlorhexidine, or isopropyl alcohol
- Sterile fluids – lactated Ringer's solution or 0.9% sodium chloride
- Oral speculum (feline) and oral feeding tube (e.g., 8 French catheter)

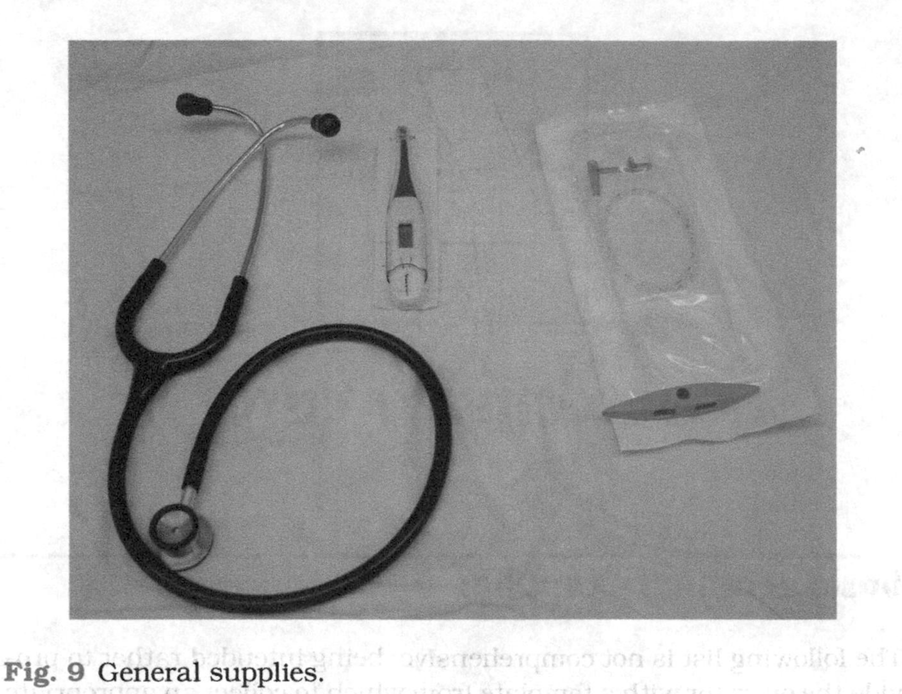

Fig. 9 General supplies.

- A gram scale (capable of measuring up to 3 kg)
- Record-keeping materials (clinical examination records, study-related documentation)

Personal Protective Equipment

- Lab coat
- Examination gloves (nitrile, latex, etc.)
- Bite-resistant gloves for restraint
- N95 respirator without one-way release valve (for influenza or similar respiratory studies) or particulate surgical mask
- Hair cover
- Eye protection

the physical examination

The proficient performance of the physical exam is one of the most important and practical skills a clinician or veterinary technician can develop. This is as true for the ferret as it is for any other experimental

Physical Exam Study number: xxxxxxx
Animal ID: 1234
Date: m/d/y Study Day: ____
Gender: Male or Female Species: Ferret **Weight (kg):** ____

General Health Status
Temp.: F **Heart Rate:** bpm **Res Rate** bpm

Body Condition Score: 1 2 3 4 5

(Scores: 1 = emaciated, 2 = thin, 3 = normal, 4 = overweight, 5 = obese)

	Normal (✓)	Abnormal (✓)	Explain any abnormalities, if present:
Hydration			
Teeth and Gums			
Superficial Lymph Nodes			
Membrane color			
Integumentary System			
Musculoskeletal System			
Respiratory System			
Cardiovascular System			
Alimentary System			
Urogenital System			
Nervous System			
Endocrine System			
Abdominal masses			

Normal and acceptable for study? (circle one): Yes / No

Clinician/Technician: ____ Date: ____

Recorder ("N/A" if clinician/Technician): ____ Date: ____

Fig. 10 Example of examination form.

animal. In the case of the ferret, it can also be particularly challenging. Ferrets are stoic animals and tend to mask clinical signs well.[42]

Physical examinations should be conducted upon arrival at the research facility, during the quarantine period, or prior to any experimental study or surgery. A good physical exam can reveal many underlying conditions and help optimize the selection of animals appropriate for use in a particular research protocol. Subtle changes in an animal's general condition or demeanor may indicate underlying conditions that may adversely impact a study. Abnormal findings detected during the physical exam must be duly documented so the clinician and investigator can determine the suitability of a particular animal for study.

In a research facility utilizing ferrets, it is not uncommon to perform physical exams on large groups of animals during a single session. Clinicians or well-trained technicians, working in pairs, with one operator providing physical restraint and the other conducting an examination and recording observations, can successfully process large numbers of ferrets quickly and easily as they gain experience.

the ferret examination

The following is a template for conducting a thorough physical examination in ferrets.

Signalment and History

- The age, sex, identification number, coat color, animal source (vendor), dietary history, and reproductive status are key pieces of information that should be collected prior to examination. These characteristics should correspond with the vendor's report when receiving animals from off-site and should be checked to ensure accuracy.
- Age may be estimated by size and body weight. Eruption of permanent teeth is not a helpful indicator for aging a ferret. By weaning (6–8 weeks postnatally), almost all permanent teeth are erupted. By day 74 of life, all deciduous teeth are shed.[43] Knowing a ferret's approximate age may help narrow differential diagnosis and aid in making decisions regarding animal management.
- Small ear tags or microchips are commonly used for ferret identification (**Figure 8**).
- The majority of the ferrets used in research have been surgically sterilized shortly after weaning. Anal sacculectomy, colloquially known as "de-scenting," is often performed concurrently at breeding facilities. For special studies, intact animals may be used. Animal colors may be used in research to identify ferrets of a particular sex or possessing characteristics of interest. For example, a vasectomized ferret used during reproductive studies may be selected with a white coat, allowing it to be easily distinguished from other ferrets in the study.[44]
- The medical history should indicate the date of birth, all vaccinations received, previous medical treatments, and surgical

procedures, including the relevant dates of service. Influenza specific-pathogen-free ferrets are usually prescreened at barrier facilities, and influenza titers should be reported and reviewed prior to enrolling an animal in an applicable study.

General Appearance/Initial Observations

- It is important to observe the ferret(s) from a distance and prior to handling. Note any asymmetry or difference in the size or shape of the extremities. Watch the ferret's gait for any signs of lameness, ataxia (uncoordinated movement), or abnormal placement of a limb that may indicate a deficit in proprioception. Normal locomotion in ferrets is characterized by arching of the back as the animal moves.[45]
- **Body weight** is a key indicator of health. As ferrets tend to hide clinical signs, the weight can reveal whether the ferret is growing according to the normal curve or whether it has recently lost body mass from disease or competitive behavior with cage-mates (e.g., a smaller ferret grouped with larger cage-mates may not have adequate access to food).
- The **body condition score (BCS)** is assigned on a five-point scale (BCS = 1–5). In general, if an animal is too thin, its ribs are easily seen and adipose tissue will not be appreciated on palpation. In an obese animal, it is difficult to palpate the ribs (**Table 5**).
- **Mentation** involves assessing the level of consciousness and reactions of the animal to the environment. A normal, conscious ferret is usually bright, alert, and responsive (BAR). It is important to remember that ferrets are (although not strictly) nocturnal animals. During the day, ferrets play in brief bursts of 15–30 minutes, after which they may sleep if not stimulated. They may sleep as much as 18 hours per

Table 5: Body Condition Score in Ferrets[53]

Body Condition Score	Comments
1	Cachectic. >20% underweight
2	Lean. 10–20% underweight
3	Normal
4	Stout. 20–40% overweight
5	Obese. >40% overweight

(From Oglesbee, B.L. 2006)

day.[46] It may be necessary to wake a ferret for examination, but once aroused they quickly engage in exploration of their environment.[46,47] This attribute may confound the evaluation of possible lethargy or depression in ferrets.

- **Hydration** may be assessed by noting the skin turgor of eyelids, the tenting of the skin at the back of the neck, and moistness of the oral mucosa (caution must be exercised to avoid being bitten; a tongue depressor may facilitate examination of the mouth). Dehydration is often underestimated in ferrets.[48] The tough skin makes the skin turgor test less reliable than in other species, especially in borderline dehydration cases. Animal activity is a better clinical indicator in such cases.[48]

Temperature, Pulse, and Respiration (TPR)

- **Temperature** Ferrets tend to be reluctant to have their temperatures taken per rectum. While the animal is being restrained by a holder, offering a food treat such as Nutrical (Vétoquinol USA, Buena, NJ) on a tongue depressor can distract the ferret long enough to successfully measure its temperature. Care must be taken to avoid damaging the rectal mucosa. A plastic digital thermometer is recommended for this procedure. Some clinicians prefer to take the temperature when starting the physical exam to avoid a stress-induced increase in body temperature. In the absence of clinical signs of disease, temperature collection may be delayed or waived in newly arrived ferrets until they acclimate to their new environment. Measurement of the rectal temperature often induces the ferret to defecate.[42] The normal temperature in the ferret is reported to be 100.0–104.0°F (37.8–40°C), with an average value of 101.9°F (38.8°C).[46]
- **Heart rate** or **pulse** is normally high in the ferret (200–400 beats per minute). The heart lies between the sixth and eighth ribs; cardiac auscultation is therefore performed in a more caudal location than in other species. Peripheral arterial pulse assessment is unreliable in the ferret.[46] Sinus arrhythmia (an intermittent decrease in heart rate) is not uncommon and may be pronounced. This should not be confused with a pathological bradycardia (slow heart rate).[49] Owing to the high heart rate, cardiac murmurs may be difficult to appreciate.

- **Respiratory rate** is determined by observation. A typical respiratory rate is 33–36 respirations per minute.[42] Normal lung sounds should be minimal.
- **Mucous membrane** (MM) color provides a quick, qualitative indication of the blood flow to peripheral tissues. Moreover, **capillary refill time** (CRT) reflects the perfusion of peripheral tissues. Pressing a finger on the gums will cause a blanching that should abate, returning to pink, less than 2 seconds after pressure is released. A prolonged CRT (>2 seconds) may indicate compromised circulation due to cold, shock, cardiovascular diseases, anemia, dehydration, use of certain anesthetics, or other etiologies that need to be evaluated.

Examination of Body Systems

- **Ears, eyes, nose, and throat** (EENT) should be inspected (along with facial symmetry)
    - **Eyes** – Cataracts can occur in both young and old ferrets. Retinal degeneration may lead to alterations in pupil dilation.
    - **Ears** – Check for signs of discharge. Brown exudates may indicate the presence of ear mites (*Otodectes cynotis*).[45,50]
    - **Nose** – Look for nasal discharge and check for sneezing. Ferrets are normally nasal breathers but can breathe orally if necessary.
    - **Throat** – The throat, teeth, and oral cavity must be inspected. Dental calculus and gingivitis are common problems in the ferret. The tonsils are elongate in the ferret and normally buried in tonsillar crypts of the soft palate.
    - **Facial symmetry** and head position should also be noted. Although infrequent, salivary mucoceles may alter the facial symmetry.[46] Oral neoplasia, such as squamous cell carcinoma, may cause bone deformation in ferrets. Head tilt is not an uncommon finding in young ferrets and may indicate a neurologic disorder or otitis media.[51]
- **Lymphatics** (LYMPH)
    - **Lymph nodes** are easily palpated (including the submandibular, prescapular, axillary, popliteal, and inguinal nodes). The nodes should be soft and may be slightly enlarged in large and overweight animals. Enlargement and asymmetry should be appropriately recorded.[52]

- Abdominal palpation of the ferret may be performed by holding the ferret off the table, scruffing it, or by supporting the animal with one hand under the thorax.[46] This restraint allows the organs to move downward and make gentle palpation of abdominal contents easy. It is not uncommon to find an enlarged **spleen**. Splenomegaly may or may not be significant but should always be noted.[53]

- **Urogenital** (UG)
  - Examine the genital area. Ferrets for research are often neutered and de-scented at a young age. However, certain types of research require intact animals.
  - To detect estrus in intact females, vasectomized males may be required.[44,52]
  - An enlarged vulva in a spayed female is consistent with adrenal disease or an ovarian remnant.[54]
  - An intact jill (female ferret) shows an enlarged vulva when in estrus. The presence of petechiae (small dark red spots) on the mucous membrane in conjunction with an enlarged vulva may indicate hyper-estrogenism and the possibility of subsequent bone marrow suppression.
  - In intact males (hobs), testes are palpable in the scrotal sac only from December to July in the northern hemisphere.[42]

- **Dermatological** (DERM)
  - It is important to check the coat for signs of alopecia (hair loss).
  - Symmetrical, bilateral alopecia, progressing from the tip of the tail cranially to the back, is a sign of adrenal disease in ferrets.
  - Scratches or alopecia on the neck or back require further investigation. Pruritus (itchiness) may indicate the presence of ectoparasites, such as fleas or mites (*Sarcoptes scabiei*), or adrenal disease.[50,53]
  - Skin scrapes, impression smears, fungal cultures, or skin biopsies may all be examined, as in other species, to help diagnose dermatological conditions.
  - Ferret claws are non-retractile and should be examined. Some authors recommend clipping long toenails to prevent accidental scratching of handlers.[42]

- **Cardiovascular and Respiratory** (CV and RESP)
  - These systems may be examined in a similar fashion to other small animals. Information on cardiac auscultation is listed under the Temperature, Pulse, and Respiration section of this chapter.
- **Gastrointestinal** (GI)
  - Clinical history (including signalment) and abdominal palpation (as described above in Lymphatics) provide valuable information for the evaluation of the gastrointestinal tract. Foreign bodies, masses, and the presence of increased intestinal fluid and gas are often appreciable with gentle palpation.

common clinical conditions in the laboratory ferret

Diarrhea

- Diarrhea is one of the more common manifestations of diseases in ferrets.[45,53,54] It seems that virtually any event disrupting normal homeostasis is capable of inducing diarrhea in this species.[55]
- Any disorder affecting the already very brief normal gastrointestinal transit time (approximately 4 hours) may lead to diarrhea.
- Normal ferret stool is tubular and formed, but only semi-solid and never hard.
- Diarrhea may vary from mucoid and green to hemorrhagic. The color of the diarrhea is not specific to a particular disease. It is often difficult to classify ferret diarrhea as large or small intestinal.
- Diarrhea may be of infectious or non-infectious etiology. Non-infectious causes include sudden changes in the diet and foreign bodies. Infectious diseases causing diarrhea include parasites (*Coccidia* spp.), bacteria (*Lawsonia, Salmonella, Campylobacter,* and *Helicobacter* spp.), and viral (rotavirus, coronavirus, and distemper virus).
- Routine stool examination, including fecal flotation, direct examination, or fecal culture, may be conducted as in other species. Common GI conditions and their associated diagnostics are listed in **Table 6**.

Table 6: Common Gastrointestinal (GI) Diseases in Ferrets[7,45,46,55,73–77]

GI Diseases	Agents	Clinical Signs	Diagnostic(s)
GI parasites	*Coccidia* *Giardia* Nematodes (rare)	Diarrhea. Weight loss. Young ferrets after weaning are often infected with coccidia.	Fecal flotation test.
Neonatal diarrhea or neonatal sickness	Rotavirus and secondary bacterial infection	Diarrhea. Stunted growth. Dehydration. Kits look wet and hairs from the head and neck are slicked down. Ferrets from 1 to 7 days old affected	Electronic microscopy of feces or intestinal cells from dead kits. Diagnosis is usually based on clinical signs. RT-PCR can detect both group A&C rotavirus. ELISA can only identify group A.
Proliferative bowel disease (PBD)	*Lawsonia intracellularis*	Chronic diarrhea. Diarrhea is profuse and contains mucus and blood. Wasting disease. Abdominal discomfort. Severe weight loss. Dehydration. Affects younger ferrets from 12 to 24 weeks old.	Biopsy or necropsy confirms diagnosis.
Inflammatory bowel disease (IBD)	Autoimmune disease Multiple causes suspected	Chronic severe diarrhea with or without mucus or blood. Hypoalbuminemia dehydration, weight loss, anorexia and vomits may occur. Ferrets older than six months are reported.	Biopsy or histopathology confirms diagnosis.
Gastritis and gastric ulcers	Severe stress (rapid growth, dietary changes, concurrent illness) and *Helicobacter mustelae*)	Often unspecific. Lethargy, abdominal discomfort, tarry stools, anorexia, weight loss. Vomits, anemia. Illness occurs in ferrets from 12 to 20 weeks old.	Fecal occult blood positive. Postmortem exam
Epizootic catarrhal enteritis (ECE)	Enteric coronavirus	Catarrhal diarrhea, dehydration, weight loss. Green and mucoid diarrhea was reported, but not pathognomonic. Young ferrets usually have mild or moderate signs or are asymptomatic. Older ferrets are more severely affected.	Biopsy or histopathology confirms diagnosis.
Campylobacteriosis and Salmonellosis	*Campylobacter*, *Salmonella* spp.	Uncommon in research facility. May occur when ferrets are fed raw meat.	Fecal culture confirms diagnosis.

Weight Loss (or Decreased Food Intake)

- Although nonspecific, these clinical signs may be related to husbandry problems, such as abrupt changes in diet or inability to obtain adequate intake because of cage-mate competition, or may be the result of underlying medical conditions.
- A physical examination should be conducted on animals exhibiting pronounced weight loss or failure to thrive. Emphasis should be placed on identifying potential underlying conditions (particularly those resulting in diarrhea or dental abnormalities that may result in the inability to properly ingest food and water).
- Care should be taken to ensure that ferrets are co-housed with animals of similar size and build. Smaller animals should be grouped together, whenever possible, to allow more uniform access to food and water.

Alopecia

- A seasonal alopecia (of unknown origin) has been reported in normal ferrets at the end of the winter and early spring.
- Hyperadrenocorticism is one of the common causes of alopecia in the ferret.[53,55,56] It occurs in male and female ferrets between two and seven years of age. This alopecia is usually symmetrical, beginning on the rump, the tail, or flanks, and eventually spreading to the sides, dorsum, and ventrum. Pruritus may occur in some cases.
- Asymmetrical causes of alopecia may be diagnosed by the methods described under the Dermatological section above. Mechanical causes of hair loss should also be ruled out.
- See **Table 8** for endocrine diseases and their associated diagnostics.

Vomition

- Ferrets are good models for human emesis, despite the fact that vomiting is often not observed or reported as being associated with many common ferret diseases.
- Vomiting must be distinguished from regurgitation (although regurgitation occurs less frequently than emesis in the ferret).[46] Regurgitation generally occurs shortly after eating, and the ingested material is often tube-shaped, unlike vomit.

- Vomiting may result from primary gastrointestinal problems (e.g., foreign bodies, gastritis, or gastroenteritis) or metabolic causes (including conditions such as kidney or liver failure).
- Regurgitation may occur with conditions that result in megaesophagus.

Respiratory Difficulty

- Difficulty breathing or other respiratory signs may be present in young ferrets with bacterial pneumonias, canine distemper, influenza viral infections, or mediastinal lymphoma (affecting the thymus in ferrets less than two years old).[18]
- In older ferrets, cardiac disease, such as dilated cardiomyopathy (DCM) should be ruled out along with respiratory infections.[49]
- Auscultation of the chest (for rales, crackles, cardiac murmurs, etc.), thoracic radiographs, EKG, echocardiography, and bacterial culture are all diagnostic modalities for respiratory conditions.
- For a list of common respiratory pathogens and their associated diagnostics, refer to **Table 7**.

Other Conditions

- **Hindlimb signs**, including ataxia (uncoordinated movement) and posterior paresis (hindlimb weakness), may result in decreased mobility. Neuromuscular disorders and systemic diseases (such as insulinoma and dilated cardiomyopathy) may also result in these clinical signs[57] (**Table 8**).
- A **swollen vulva** is a common sign of estrus in intact females, but may also be a clinical sign of adrenal disease or an indication of a remnant ovary in spayed jills.[55] Alopecia is often present. **Anemia** may also be present with either disease (adrenal tumor or remnant ovary) and may be caused by endogenous estrogen toxicity (**Table 8**).

preventive medicine

A preventive medicine program is vital to the success of research facilities utilizing ferrets. Such programs commonly include health assessment and screening, vaccination, parasite control (for internal and external parasites), and quarantine procedures.

TABLE 7: COMMON RESPIRATORY (RESP) DISEASES IN FERRETS[7,42,73,78,79]

Disease	Agent	Clinical signs	Diagnostic(s)
Distemper	Canine distemper virus Morbilivirus	Fever, nasal and ocular discharge (mucopurulent), depression, anorexia. Footpads and nose hyperkeratosis. Secondary bacterial pneumonia with cough. CNS signs. Vomiting and diarrhea may also occur. Almost 100% mortality.	Clinical signs and postmortem diagnosis: immunochemistry, histopathology and immunofluorescence, virus isolation, PCR. Inclusion bodies in brain, lungs, bladder, stomach, and lymph nodes.
Influenza	Human influenza virus (varied strains)	Fever and mucoserous discharge, coughing, sneezing as main clinical signs. Upper respiratory signs. Self-limiting.	Clinical signs, virus isolation from nasal discharge and antibody titer.
Bacterial pneumonia	Gram+ (*Streptococcus* spp.) and Gram– (*E. coli, Klebsiella, Pseudomonas aeruginosa*)	Lower respiratory tract signs. Not common in ferrets. Bacterial pneumonia often occurs secondarily to viral diseases such as influenza or CDV.	Clinical signs, radiographs. CBC and microbial cultures from tracheal wash. Postmortem confirmation by histopathology and bacterial culture.
Pulmonary mycoses	*Blastomyces dermatitides Coccidiomycosis immitis*	Unlikely in ferrets kept indoors in research facilities. Lower respiratory signs.	Diagnosis based on cytologic identification of the agents and history of travel or location of animal in areas where etiologic agent is endemic.

Health Assessment and Screening

- The proper assessment of the health status of incoming animals is paramount to the success of a preventive medicine program. Proper physical examination technique and selection of appropriate diagnostics (as described above) are imperative.
- Ferrets are experimental models for human influenza research. Flu-free ferrets are often screened for serum antibody titers to various human flu strains at the vendor facilities

Table 8: Common Endocrine (ENDO) Diseases in Ferrets[50,53,55,56,73,78]

Disease	Etiology	Clinical signs	Diagnosis
Insulinoma	Functional neoplasia of pancreatic beta cells	Affected ferrets range in age from 3 to 9 years. Main signs include disorientation, hind limb weakness, and collapse. Hyper salivation is not uncommon. Hypoglycemic episodes may be precipitated by previous exercise or fasting. Seizures may occur.	Clinical signs and demonstration of fasting hypoglycemia. Exploratory laparotomy or necropsy confirms diagnosis.
Adrenal gland disease (AGD)	Adrenocortical hyperplasia adenoma, adenocarcinoma	Reported in ferrets from 2 to 5 years old. Bilateral symmetric alopecia from the back and tail. Sexual hormone levels are high in this disease. Females may show vulvar enlargement and males may resume sexual behavior (e.g., mounting).	Clinical signs and elevated serum levels of sex hormones. Enlarged adrenal gland may be detected by palpation, ultrasound, or exploratory laparotomy.
Hyperestrogenism	Spayed female with remnant ovary Intact female in estrus	Ferrets are induced ovulators. During the breeding season estrogen levels increase, and if jills are kept unbred for more than one month they may develop bone marrow suppression due to estrogen blood levels. Alopecia, anemia, swollen vulva, and thrombocytopenia are reported.	Differential diagnosis should be done between AGD and remnant ovary in spayed jills. Alopecia and enlarged vulva are common signs in both disorders.

Table 9: Common Hemolymphatic (LYMPH) Diseases in Ferrets[18,53,73,78,79]

Disease	Etiology	Clinical Signs	Diagnosis
Lymphosarcoma	Neoplasia of unknown etiology Retrovirus suspected	Juvenile forms in ferrets <2 years of age, acute clinical signs (dyspnea, pleural effusion; caused by presence of mediastinal masses). Older ferrets have a chronic course of disease involving spleen, liver, and lymph nodes. Other atypical forms are reported with low incidence.	Confirmation by postmortem exam or biopsy of the abnormal tissue.
Splenomegaly (splenic enlargement)	Lymphosarcoma (neoplasia) Extramedullary hematopoiesis (diffuse enlargement) Other: splenitis, congestion, and lymphoid hyperplasia are uncommon.	Can cause abdominal enlargement and discomfort. Lymphosarcoma is a wasting disease (EMH). May be found in clinically normal ferrets or with no related illness (insulinoma or adrenal disease).	Biopsy or postmortem macro- and microscopic exam establishes differential diagnosis. Peripheral blood examination and differential may aid in final diagnosis.

and prior to enrollment in an experimental study. Workers at research facilities using ferrets on such studies wear special personal protective equipment (PPE), such as N95 respirators without release valves, to prevent not only transmission of an agent to themselves, but also contamination of study animals by undesired viral strains such as seasonal influenza potentially carried by animal workers (**Figure 4**). It may be necessary under certain circumstances to provide additional screening services to ensure that a study is not potentially confounded, should a breach in protective practices occur.

Vaccination

- Ferrets are routinely vaccinated for canine distemper virus (CDV) and rabies. CDV vaccination is imperative, as maternal immunity wanes by six weeks of age and the disease can produce close to 100% mortality in mustelids such as ferrets. A serial vaccination protocol is recommended, with vaccination at

6–8 weeks, 10–12 weeks, and 13–14 weeks of age sequentially. Thereafter, an annual booster vaccination is recommended.

- Vendor facilities may use the mink distemper vaccine in their ferrets. This is often given as a polyvalent vaccine, protecting against CDV, mink enteritis virus (MEV), *Pseudomonas aeruginosa* pneumonia (PAP), and *Clostridium botulinum*. PAP and MEV are mink-specific agents and, as such, are not medically indicated in research ferrets. Clinicians and investigators should be aware (particularly if they are studying vaccination against respiratory agents) that a monovalent CDV vaccine may be preferred under such circumstances.
- Rabies vaccination is recommended at 12 weeks of age, and annually thereafter. The decision to vaccinate in a research environment depends on facility type (indoor or outdoor) and the associated environmental risk (rabies-endemic area). The IMRAB 3 (Merial, Athens, GA) vaccine is approved for use in ferrets and should be administered annually.
- Annual influenza vaccination is recommended for personnel working with ferrets enrolled in influenza studies.

Parasite Control

- Coccidia (*Isospora* and *Eimeria* spp.) are the more commonly found endoparasites in ferrets (especially young animals). If a program of random fecal direct and flotation examinations on quarantined animals detects such organisms, treatment is indicated. Treatment options for coccidiosis are listed in **Table 11**.

Table 10: Other Diseases in Ferrets[73,78-81]

Disease	Etiology	Clinical Signs	Diagnosis
Aleutian disease	Parvovirus	Wasting disease. Unspecific signs. Hypergammaglobulinemia and autoimmune glomerulonephritis at end stages is reported.	Serology, hyperproteinemia. Postmortem lesions to confirm the disease.
Systemic coronavirosis	Systemic coronavirus	Wasting disease. Unspecific signs. Pyogranulomas on liver and spleen on postmortem exam. Masses may be palpated antemortem. Similar lesions to feline infectious peritonitis (FIP) dry form.	Postmortem macroscopic and microscopic lesions confirm the disease (pyogranulomas).

Table 11: Common Parasiticides for Ferrets[45,76,82-84]

Parasiticide	Route	Dosage	Comments
Ivermectin	SC, Topical	0.2 mg/kg SC every 7–14 days	For external parasites (*S. scaibei* and *Demodex* sp.). For ear mites instilled every 14 days.
Imidiclorpride	Topical	0.1 mL monthly	For fleas.
Metronidazole	PO	15–25 mg/kg bid for 5 days	For *Giardia* sp. Due to a bitter taste ferrets can vomit or salivate. It is effective against *Helicobacter m.*
Praziquantel	PO, SC	5–10 mg/kg repeat in two weeks	For cestodes.
Pyrantel pamoate	PO	4.4 mg/kg repeat in two weeks	For nematodes.
Sulfadimetoxine	PO	50 mg/kg first day and 25 mg/kg sid for 9 days.	For coccidia.
Selamectin	Topical	6–12 mg/kg once monthly (Oglesbee, 2006)	Effective against ear mites. Same dose as indicated in the cat (author personal experience). Not recommended for ferrets younger than two months.
Trimethoprime-sulfa combination	PO	30 mg/kg q 24 h for two weeks	For coccidia.
Ponazuril	PO	30 mg/kg. Repeat in one week.	The dose is based on safety and efficacy studies of 5% suspension performed on dogs.

SC, subcutaneous; PO, oral (per os); sid, once a day

- Ear mites (*Otodectis cynotis*) may be encountered in a research setting. Topical and/or injectable drugs are used to combat these and other ectoparasites. See **Table 11** for more information.

Quarantine and Acclimation

- When ferrets are received into a facility, they are often quarantined in an area physically isolated from the general animal population. A complete physical exam should be performed upon arrival or shortly thereafter with particular emphasis on ruling out possible communicable conditions (respiratory or other viral or bacterial infections of interest).

- The quarantine period should be long enough to permit daily clinical observation (including body weight changes and food and water consumption patterns) and to allow reasonable certainty that newly arrived animals are free of conditions that may adversely affect colony health. A period of one to two weeks is usually required to identify ferrets with respiratory or intestinal pathogens.[52]
- Animals showing clinical signs of illness should be separated from the remainder of the group for veterinary care or culled as appropriate.
- Ferrets require acclimation to new diets and new watering devices (e.g., automated watering systems). The transition must be done gradually, phasing in the new diet (if applicable) and providing supplemental water as necessary until the ferrets adjust to the new system.
- Collecting feces from randomly selected cages allows screening for various gastrointestinal parasites (e.g., Coccidia).

therapeutics

It is important to be mindful that drugs commonly used in ferrets are not labeled for the species. There is a dearth of pharmacokinetic toxicity information on many therapeutic agents used in ferrets. Drug dosages and dosing intervals are often extrapolated from those used for the dog or cat, or from an individual use as reported in a publication, or is based on the clinician's personal experience.

When considering the use of therapeutic agents in ferrets, it is important to remember that while dogs, cats, and ferrets have many similarities, there are also some significant differences. Ferrets, like dogs and cats, have a strong vomit response, but the ferret's gastrointestinal transit time is markedly shorter than that of either species.[46] The decreased transit time may affect total compound absorption (whether the drug is given clinically or experimentally). This is a single example of a species difference with profound implications that must be kept in mind when medicating ferrets.

Drugs can be administered orally or parenterally in ferrets. Tablets are not recommended for oral dosing of ferrets, because of the ferret's small mouth and its tendency to bite during administration. Tablets or capsules may be crushed or suspended into a vitamin B supplement paste (or similar appropriate solution) for oral administration.

Liquid suspensions or oral drops can be instilled inside of the mouth with a small syringe or a dropper. Ferrets do not have a strong cough reflex, and if a ferret struggles during oral administration, aspiration pneumonia may occur. Several medications (e.g., metronidazole, prednisone, and bismuth subsalicylate) have a very bitter taste and may exacerbate the ferret's resistance to oral administration. To minimize ferret stress when using such medications, the use of a compounding pharmacy to provide a flavored vehicle for the compound is invaluable.

Subcutaneous injections are given between the shoulder blades. Fluids may be given, subcutaneously superficial to the ribs or on either side of the lumbar spine. Ferrets lack significant muscular mass. Intramuscular injections are given into the quadriceps muscle. Intravenous injections are usually performed on sedated or anesthetized ferrets. The cephalic vein is the most commonly used vessel for peripheral venous access.[58]

It is vital that veterinarians discuss therapeutic options with the investigator prior to their use in order to avoid introducing agents that may interfere with or confound interpretation of relevant study data (**Tables 11–13**).

anesthesia and analgesia

Definitions

- **General anesthesia** is the controlled loss of sensation to pain with unconsciousness and muscle relaxation.
- **Sedation** is the depression of consciousness while maintaining voluntary muscle movement without analgesia.
- **Preemptive analgesia** is the administration of analgesics in anticipation of rather than in response to surgical pain.
- **Anesthesia induction** in general anesthesia is a transition period from a conscious state to an unconsciousness state or anesthetic level.
- **Anesthetic period** is the time of effective anesthesia used to perform a procedure. During this period, anesthesia may be light, adequate, or deep.
- **Recovery period** is the time interval when the patient's brain concentration of anesthetic drops below a certain key level. This is the period between the end of an experimental or

Table 12: Common Antibiotics for Ferrets[76,85]

Antibiotic	Route	Dosage	Comments
Amoxicillin	PO, SC	10–35 mg/kg bid	Can result in decreased appetite. Used for ulcerative gastritis with metronidazole and Pepto-Bismol (bismuth subsalicylate) to eradicate *Helicobacter mustelae.*
Amoxicillin-clavulanic acid	PO	10–20 mg/kg bid or tid	General antibiotic therapy.
Clarithromycin	PO	50 mg/kg sid	Effective against *Helicobacter mustelae* combined with ranitidine and bismuth citrate.
Clindamycin	PO	5–10 mg/kg bid	General antibiotic therapy.
Cephalexin	PO	15–25 mg/kg bid to tid	Can potentiate aminoglycoside toxicity.
Chloramfenicol	PO, SC	50 mg/kg bid	To treat proliferative bowel disease (PBD) Ineffective against *Helicobacter m.*
Gentamicin	SC	4 mg/kg sid	Potentially nephrotoxic.
Neomycin	PO	10–20 mg/kg bid	Treatment of diarrhea.
Penicillin procaine	IM	20,000–40,000 UI/kg sid	General antibiotic therapy.
Tylosin	PO	10 mg/kg sid or bid	Treatment of diarrhea.

PO, oral (per os); SC, subcutaneous; IM, intramuscular; sid, once a day; bid, twice a day

clinical procedure and the animal regaining the righting reflex or ability to maintain its posture in sternal recumbency (lying upright on its stomach with control of its head and airway).

Anesthesia

- Anesthesia is ideally maintained at an adequate level (the "surgical plane") necessary to perform the planned procedure. In general, inhalant anesthetics are more suitable for controlling the anesthetic level or "plane" than are injectable anesthetics. The anesthetic level of most injectable anesthetics cannot be controlled once they are administered by intramuscular or intraperitoneal routes. Intravenous agents can, under certain circumstances, be used to provide additional control over the anesthetic level when they are administrated by continuous rate infusion (CRI) and in very specific dose rates.
- Commonly used anesthetics in ferrets include isoflurane (an inhalant anesthetic) and ketamine/xylazine/atropine (a combined injectable anesthetic cocktail).

Table 13: Miscellaneous Drugs in Ferrets

Drug	Route	Dosage	Comments
B-vitamin complex	PO, SC	1–2 mg/kg sid	Appetite stimulant. For debilitated ferrets.
Bismuth subsalycilate	PO	0.25 mg/kg tid	Bitter drug strongly resisted by ferrets.
Cimetidine	PO, SC	5–10 mg/kg tid	H2 receptor antagonist. For gastric ulcers. It has been replaced by ranitidine (3.5 mg/kg bid PO).
Diphenylhydramine	PO, IM	2 mg/kg sid	Prevaccination dose for sensitive ferrets.
Famotidine	PO, IM	0.25–0.5 mg/kg sid	H2 receptor antagonist. For gastric ulcers.
Nutrical	PO	2–5 mL bid	Multiple nutritional supplement in paste used in debilitated ferrets.
Oxytocin	SC, IM	0.2–3.0 IU/kg	For parturition induction with Prostaglandin F_2 alpha.
Phenobarbital	PO	1–2 mg/kg q 8–12 hs for seizure control. Titrate dose for maintenance	Anticonvulsive therapy.
Prednisone	PO, SC	0.5–2.5 mg/kg sid or bid	Steroid anti-inflammatory.
Prostaglandin F2-alpha	SC	0.5 mg/kg	For parturition induction with oxytocin.
Sucralfate	PO	10–20 mg/kg q 4–6 hours	For gastric ulcer treatment.

PO, oral (per os); SC, subcutaneous; IM, intramuscular; sid, once a day; bid, twice a day; tid, three times a day

Sedatives

- **Acepromazine** is a tranquilizer. It does not provide analgesia and is, therefore, inappropriate for single-agent use when performing surgical interventions. Ferrets receiving acepromazine at 0.1 mg/kg IM are generally sedated for 40–50 minutes.[59] Higher doses (0.2–0.5 mg/kg) prolong the recovery time, but lower doses do not have sedative effects.[59] Caution should be exercised when using acepromazine, as it produces profound vasodilatation and may lead to hypotension and hypothermia, especially with high doses and in dehydrated ferrets.
- Tranquilizers, such as **diazepam** and **midazolam**, are not recommended as sole agents for sedation of ferrets. Problems associated with intramuscular injections and lack of consistency

in level of sedation and muscular relaxation produced preclude their use as such.[59] They also do not provide analgesia, although they do reduce anxiety.

Injectable Anesthetics

- Alpha agonists, such as **xylazine**, **medetomidine**, and **dexmedetomidine**, provide profound sedation, muscular relaxation, and analgesia. Side effects include hypotension, bradycardia, and respiratory depression. They have an advantage over other injectable agents used in ferrets, in that the pharmacological effects of these drugs can be reversed with alpha antagonists, such as yohimbine or atipemazole.[59,60]

Pre-Anesthetic Preparation

- **Fasting** ferrets, like other domestic carnivores, have a very strong emetic response (urge to vomit) when given certain medications, including anesthetics.[42] Ferrets also have a very short gastrointestinal transit time (4 hours) compared to dogs and cats.[61] Additionally, ferrets more than two years old are prone to developing insulinomas that may lead to hypoglycemia.[45,56,57] The combination of these factors dictates that a ferret requiring anesthesia and surgery should be fasted, but for a shorter duration relative to dogs and cats (8–12 hours). In the literature, many fasting times have been proposed by various authors. These tend to break down into two general approaches, with a maximum fast of either 8–12 hours or 3–4 hours.[42,59,62-65] Fasting periods of no longer than 8 hours for young ferrets and 4 hours for middle-aged or geriatric ferrets are proposed. The rationale for the varying fast times by age relates to the risk of hypoglycemia secondary to possible insulinoma.[65] In the authors' experience, 2–4 hours is a reasonable fasting time for ferrets. Prolonged fasting times may induce a hypoglycemic state and complicate anesthetic recovery. Animals should, however, be allowed ad libitum access to water until anesthesia or one hour before. In order to maintain normoglycemia, 5% dextrose in water may be administered during the preoperative or intraoperative periods.

- **Physical examination and preoperative blood work**. A complete physical examination is recommended prior to any experimental procedure involving anesthesia and surgery in the ferret. A minimal blood work database may be collected in a healthy ferret. This should include packed cell volume (PCV), total protein (TP), blood glucose (BG), and blood urea nitrogen (BUN). A laboratory animal veterinarian should be consulted when developing an anesthetic protocol suitable for the research project.
- **Premedication**. Various types of agents may be used as premedication in ferrets.
    - The **anticholinergics** atropine and glycopyrrolate are used to prevent bradycardia (slow heart rate) and reduce salivary secretions in the airway precipitated by the use of other drugs (such as ketamine). Both agents reduce respiratory secretions; however, care must be exercised as they may thicken airway secretions and cause obstructions.[66]
    - **Acepromazine** or **alpha-2 agonists** (xylazine, dexmedetomidine) may be administered as premedication, or for induction when combined with ketamine, a dissociative anesthetic.
    - Preemptive analgesia may be achieved with **opioid agonists**, such as **morphine**, **oxymorphone**, **hydroxymorphone**, and **mixed agonist-antagonists** such as **buprenorphine** and **butorphanol**. Alpha agonists exert analgesic activity with sedative effect.
    - **Nonsteroidal anti-inflammatories** (NSAIDs) that lack sedative effects, such as **flunixin meglumine**, **carprofen**, **ketoprofen**, and **meloxicam**, are used for preemptive analgesia.
- **Preoperative fluids**. Ferrets may benefit from preoperative and intraoperative fluid supplementation. Usually, 10 mL/kg preoperatively will alleviate dehydration in preparation for surgery. Ideally, intravenous fluids are preferable, but tough skin and large amounts of subcutaneous fat render IV catheterization difficult.[58] However, clinically normal ferrets absorb subcutaneous fluids well, and up to 20 mL/kg may safely be given via that route, or up to 10 mL/kg IV may be given slowly. Lactated Ringer's solution (LRS) is commonly used, and dextrose in saline (2.5% or 5%) is the choice to prevent hypoglycemia in ferrets.

Anesthetic Induction

- Anesthetic induction can be achieved in various ways in the ferret. When using inhalant agents (isoflurane, sevoflurane), the procedure may be performed using an induction chamber (**Figure 11**) or a face mask (**Figure 12**).
- Excessive salivation has been reported in ferrets when using isoflurane for induction.[59] Anticholinergic agents are used to control salivation.
- Induction chambers are very useful (especially for fractious animals), but can expose operators in the room to anesthetic gas when open compared with a facial mask system. Caution must be exercised when using such a system.
- Facial masks should cover the nose and mouth. A non-rebreathing circuit (modified Jackson Reed) is connected to the mask and the anesthesia machine. High oxygen flow rate and high inhalant concentration are used for induction compared with the maintenance and anesthetic period. A face mask can also be used during the maintenance period for very quick procedures.
- Once induction is achieved, endotracheal intubation (ET) may be performed to deliver the anesthetic. For intubation, a 3-mm-internal-diameter cuffed tube is used for ferrets weighing more than 1.0 kg. For smaller animals, an uncuffed endotracheal tube may be used (**Figure 12**). The intubation technique in ferrets is very similar to the technique used for cats. Various types of laryngoscopes with a straight blade may be used. Ferrets are positioned in dorsal[63] or sternal recumbency (**Figure 13**) for the procedure. Ferrets are not as prone as cats to laryngospasm.
- Dissociative injectable combinations are administered by the subcutaneous or intramuscular routes. Ketamine can be combined with acepromazine, xylazine, dexmedetomidine, diazepam, or midazolam. Each one of the combinations has pharmacological properties useful for different experimental procedures. Tiletamine with zolazepam (Telazol) is another injectable dissociative combination. Other drugs such as central analgesics (opioids) may be part of the drug combination. After the combined drugs are injected, the ferret should be placed in a quiet environment and continuously observed.

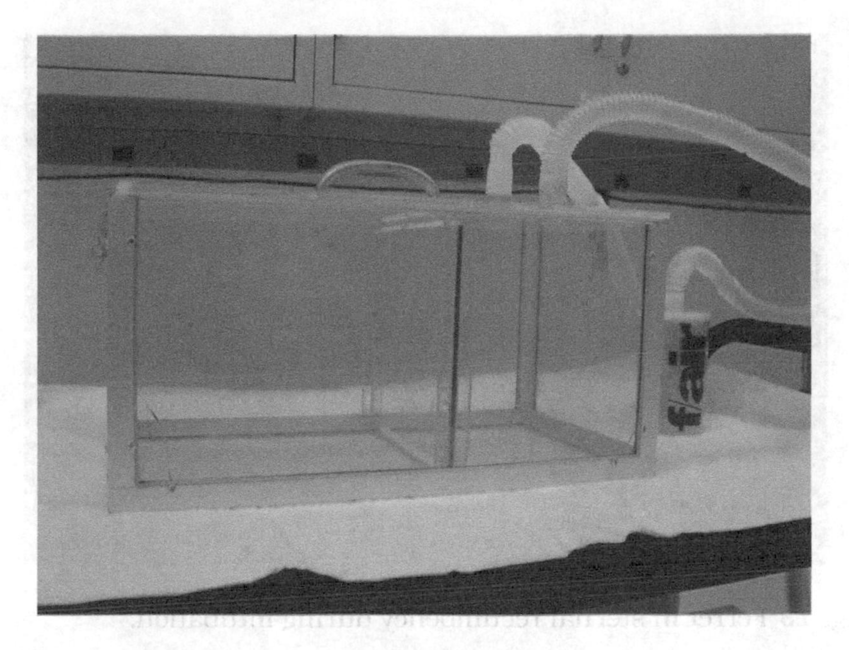

Fig. 11 Induction chamber for anesthesia.

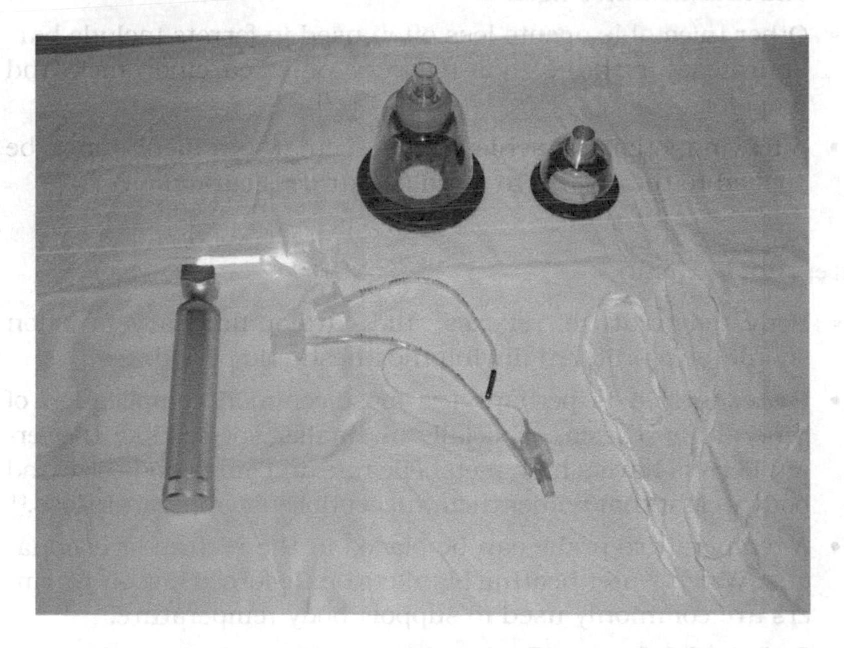

Fig. 12 Face mask and endotracheal tube for anesthesia.

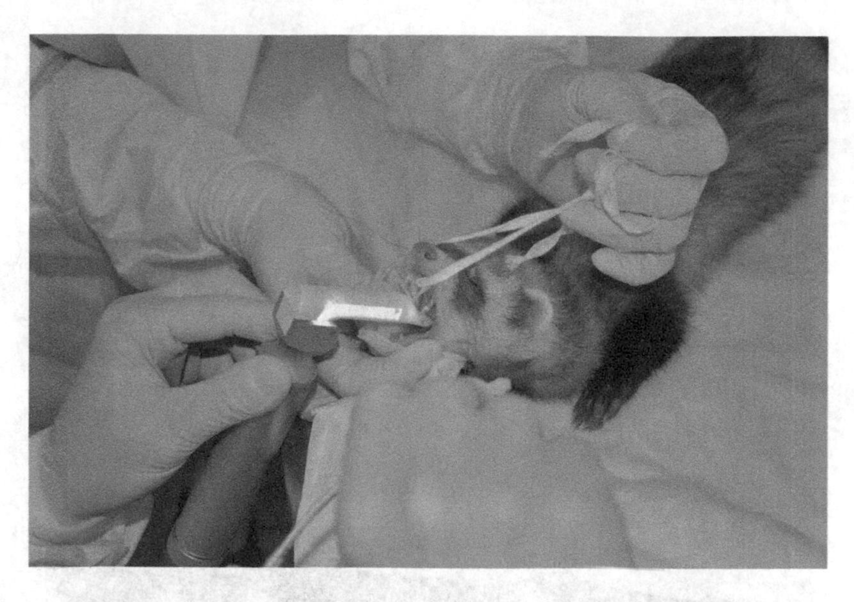

Fig. 13 Ferret in sternal recumbency during intubation.

Dissociative injectable combinations serve as both induction and maintenance agents.

- Other injectable agents less often used in ferrets include barbiturates, urethane (non-recovery only; carcinogenic), and propofol.
- After induction is achieved, an ocular ointment must be applied to the cornea to avoid ocular desiccation.

Anesthetic Period

- Body temperature, reflexes, and cardiopulmonary function should be monitored during the anesthetic period.
- **Hypothermia** is perhaps the most common complication of general anesthesia, especially in smaller species like the ferret. Ferrets have a high metabolic rate and small body size and body weight that render them susceptible to temperature loss.[66]
- A temperature probe can be placed in the rectum or esophagus. Warm-water heating blankets or air-forced hot-air warmers are commonly used to support body temperature.
- **Pedal withdraw reflex**, in conjunction with muscular tone and loss of **righting reflex**, provides a good indicator of anesthetic plane.

- Ferrets should receive the same **cardiopulmonary monitoring** as used in dogs and cats. Routine monitoring of ferrets consists of heart rate and rhythm as well respiratory rate and character.
  - Devices used to monitor cats generally work well in ferrets. Pulse oximeters have been adapted for use in ferrets. An SPO_2 reading lower than 90% is indication of hypoxia in ferrets.
  - Esophageal stethoscopes are good options for monitoring heart rate and respiratory sounds.
  - EKG monitors may also be used in ferrets.
- The American College of Veterinary Anesthesiologists has made additional monitoring recommendations applicable to ferrets.[67]

Recovery Period

Considerations for anesthetic recovery are similar to those for other species. Monitoring clinical signs and controlling hypothermia, hypoglycemia, and pain are keys for a successful outcome. *Never* leave a ferret unattended during the recovery phase. Keep in mind that pain assessment is difficult in ferrets because they tend to mask pain well. Preemptive and multimodal analgesia work well in ferrets. The recovery period is frequently longer with injectable agents and shorter with inhalant anesthetics (**Tables 14–19**).

aseptic surgery

The ferret is a USDA-regulated animal species. Aseptic technique for survival procedures is a requirement for experimental surgery in this species.[18] The goal of aseptic surgery is to reduce microbial contamination to the lowest practical level. In general, aseptic techniques include the following.

- Patient Preparation
  - Ferret preparation includes hair removal and preparation of the operation site.
    - The site is cleaned by alternating a surgical disinfectant (e.g., povidone-iodine scrub or chlorhexidine scrub) with isopropyl-alcohol-saturated gauze 4 x 4's.
    - Three alternations of scrub followed by isopropyl are performed, each starting at the center of the surgical

Table 14: Common Sedatives for Ferrets[57,58]

Drug	Route	Dosage	Comments
Acepromazine	IM	0.1 mg/kg	Sedation for minor procedures. Do not provide analgesia. Duration: < 1 hour.
Xylazine	IM, SC	2.0 mg/kg	Sedation and analgesia for minor procedures. Effect can be reversed with yohimbine (0.5 mg/kg IM) or atipemazole (1.0 mg/kg IM).
Medetomidine	IM, SC	0.02–0.04 mg/kg	Sedation and analgesia for minor procedures. Effect can be reversed with atipemazole (1.0 mg/kg IM).
Dexmedetomidine	IM, SC	0.01–0.03 mg/kg	Similar to medetomidine.

PO, oral (per os); SC, subcutaneous; IM, intramuscular

field and spiraling gradually outward to the margin of the field. Care is taken to ensure that gauze that has touched hair at the margin of the field is not pulled back toward the center (consult a surgical text[68] or laboratory animal veterinarian for further information).

- Preparation of the Surgeon and Instruments
    - The surgeon and assistant should don caps, surgical masks, and sterile gowns and gloves. A proper surgical scrub is mandatory.
    - Sterilized surgical instruments, supplies, and implanted devices must be used.
- Good Surgical Practice
    - Appropriate surgical technique, including gentle tissue-handling, helps reduce postoperative complications, including infection.
- Aseptic technique in ferrets is similar to aseptic technique in dogs and cats. Detailed procedures are covered by small-animal surgical nursing textbooks.[68]

Surgery in Ferrets—Additional Notes

From a surgical perspective, ferrets have a thick subcutis and thin abdominal musculature. The linea alba is well defined. Old ferrets

Table 15: Common Analgesics in Ferrets[86]

Drug	Route	Dosage	Comments
Morphine	IM, SC, IV	0.25–1 mg/kg Every 2–6 hours	For mild to severe pain. Bradycardia and vomition may occur with doses higher than 0.5 mg.
Hydroxymorphone	IM, SC, IV	0.025–0.1 mg/kg every 2 hours	For mild to severe pain. Bradycardia, vomition, or respiratory depression occasionally occurs. Sedative effects are not seen when used as a single drug.
Oxymorphone	IM, SC	0.05–0.2 mg/kg every 6–12 hours	Mild to severe pain.
Butorphanol	SC, IM	0.4 mg/kg Every 4 hours	Mild to moderate pain.
Buprenorphine	SC, IM	0.01–0.02 mg/kg every 8–12 hours	Mild to moderate pain.
Flunixin	SC	0.5–2.0 mg/kg every 24 hours	Non-steroidal anti-inflammatory drug (NSAID). Do not administer for a period exceeding three consecutive days. NSAID may potentially produce prolonged bleeding time or gastric ulcers. May be used for perioperative period with or without opioids for balanced anesthesia.
Carprofen	SC, IM	2.0–4.0 mg/kg every 24 hours	NSAID may potentially produce prolonged bleeding time or gastric ulcers. May be used for perioperative period with or without opioids for balanced anesthesia.
Ketoprofen	SC, IM	1.0–2.0 mg/kg every 24 hours	NSAID may potentially produce prolonged bleeding time or gastric ulcers. May be used for perioperative period with or without opioids for balanced anesthesia.
Meloxicam	SC, IM, or PO	0.2 mg/kg every 24 hours	NSAID may potentially produce prolonged bleeding time or gastric ulcers. May be used for perioperative use with or without opioids for balanced anesthesia.

(Continued)

Table 15: (Continued) Common Analgesics in Ferrets[86]

Drug	Route	Dosage	Comments
Lidocaine	Local infiltration	2.0 mg/kg (max)	Alleviates pain locally.
Bupivacaine	Local infiltration	1.0 mg/kg (max)	Alleviates pain locally.
Mepivacaine	Local infiltration	2.0 mg/kg (max)	Alleviates pain locally.

PO, oral (per os); SC, subcutaneous; IM, intramuscular

may have significant intra-abdominal fat deposits that complicate identification of intra-abdominal structures such as adrenal glands, pancreas, ovaries, ureters, and lymph nodes.

Gonadectomy and anal sacculectomy are commonly performed at vendor facilities. Other therapeutic surgical procedures (common in exotic animal practice), such as adrenal gland surgery, pancreatic surgery, and splenectomy, are not commonly performed in a research setting. Other experimental procedures, such as vessel catheterization and venous access port implantation, are occasionally performed in support of ferret research protocols. Therapeutic and experimental techniques are covered at length by various authors[44,46,69,70] and are beyond the scope of this book.

Postsurgical Care

After anesthetic recovery, ferrets require monitoring. When normal drinking and eating behavior and physiological parameters have returned, the ferret can be moved to a more standardized husbandry location.

In addition to general clinical monitoring, the incision site(s) will require attention. Sutures may need to be removed or in-dwelling catheters flushed. The wound should be observed for signs of infection (redness, swelling, heat, purulent discharge, pain on touch, and loss of function). Other complications such as fluid accumulation (serum, blood, effusion) at surgical sites, incisional dehiscence, and self-trauma may occur.

An adequate pain management protocol should be in place *prior to surgery* and extended through the postoperative period. The same drugs mentioned in the preemptive analgesia sections above (opioids, local anesthetics, and non-steroidal anti-inflammatories) are often administered for several days postoperatively to provide adequate relief and allow the ferret to return to a normal physiological state as

Table 16: Dissociative Anesthetic Combinations for Ferret Anesthesia[58,59,87]

Drug	Route	Dosage	Comments
Ketamine-Xylazine (K-X)	IM, IP, SC	K: 20–25 mg/kg X: 2.0–2.5 mg/kg K and X mix (combine 10 mL of ketamine 10% and 1.5 mL of xylazine 10%); dose at 0.25 mL/kg. (Calculated dose: K: 21.2 mg/kg and X: 3.2 mg/kg)	K-X induces adequate analgesia, muscle relaxation duration, and smooth recovery. Ventricular premature contractions (VPC), sinus bradycardia, and AV heart block may be seen with K-X. Respiratory depression RR, bradycardia hypotension, and hypothermia are side effects related to xylazine. Atipemazole or yohimbine may be used to reverse xylazine. K-X doses (K: 30 mg/kg; X: 3 mg/kg) are used for ET intubation but atropine should be used as premedication with this combination.
Ketamine-Medetomidine (K-M)	IM, SC	K: 5 mg/kg M: 80 mcg/kg	K-M induces adequate analgesia, muscle relaxation duration, and smooth recovery. Same side effects as xylazine. M effects can be reversed with atipemazole. Dexmedetomidine is an R enantiomer of M and commercially available. Has similar outcomes clinically as those observed with M.
Ketamine-Acepromazine (K-A)	SC, IM	K: 30 mg/kg ACE: 0.3 mg/kg	Provides mild anesthesia with superficial analgesia for minor procedures such as blood draw or IV catheterization. Hypothermic and hypotensive effects.
Ketamine-Midazolam (K-M)	SC, IM	K: 15mg/kg Mid: 0.4 mg/kg	Provides mild anesthesia with superficial analgesia for blood sampling and IV catheterization. M is water soluble and has better absorption than diazepam.

(Continued)

Table 16: (Continued) Dissociative Anesthetic Combinations for Ferret Anesthesia[58,59,87]

Drug	Route	Dosage	Comments
Tiletamine-Zolazepam (Telazol)	IM	12–22 mg/kg	Provides adequate anesthesia for short and minor procedures. When used alone, recovery may be prolonged and rough. It can be used alone or in combination with low doses of ketamine and xylazine. However, hypoxemia is not uncommon when used alone or in combination.
Ketamine-Xylazine-Butorphanol (BUT)	IM	K: 15 mg/kg X: 2 mg/kg BUT: 0.2 mg/kg	Irregular heartbeats and low blood oxygen. Some authors recommend oxygen delivered by mask. Buprenorphine can replace BUT in the mix at the lower dose.

IP, intraperitoneal; SC, subcutaneous; IM, intramuscular

Table 17: Inhalant Agents for Ferret Anesthesia[63,88-91]

Agent	Route	Dosage	Comments
Isoflurane (ISO)	Inhalation	Induce at 3–4% MAC: 1.52% Maintenance: 1–1.5 MAC O_2 flow: 2 L/min (non-re-breathing circuit)	Decrease in systolic blood pressure, heart rate, and hematocrit. Mask or chamber induction may be used. Excessive salivation has been reported in ferrets during induction using induction chamber.
Sevoflurane (SEVO)	Inhalation	Induce at 6.5% MAC: 2.58% Maintenance: 1–1.5 MAC O_2 flow: 2 L/min (non-re-breathing circuit)	Decrease in systolic blood pressure, heart rate, and hematocrit. Lower blood solubility results in more rapid induction and recovery compared with ISO. Mask or chamber induction may be used.

quickly as possible. A veterinarian should be consulted when developing a postoperative analgesia schedule to help assess its adequacy with respect to the surgical procedure being performed.

Depending on the animal model created, a long-term postoperative care plan may be necessary and may involve special diets, daily medication, or specialized treatments.

TABLE 18: ANTICHOLINERGICS, REVERSAL AGENTS AND EMERGENCY DRUGS IN FERRETS[59,60]

Drug	Route	Dosage	Comments
Atropine (ATR)	SC, IM	0.05 mg/kg	May be used to reverse bradycardia. Excessive salivation produced by certain anesthetic agents.
Glycopyrrolate (GLYCOP)	SC, IM	0.01 mg/kg	Same effects than atropine. GLYCOP is less arrhythmogenic and has longer duration in the GI than ATR. GLYCOP does not cross placental and blood-brain barriers.
Yohimbine	SC, IM	0.5 mg/kg	Reversal agent for xylazine.
Atipemazole	SC, IM	1 mg/kg	Reversal agent for xylazine, medetomidine, and dexmedetomidine.
Naloxone	SC, IM, IV	0.01–0.03 mg/kg	Opioid reversal.
Doxapram	IV, IM	5–10 mg/kg	Respiratory stimulant.
Epinephrine 1:10,000	IV, Intratracheal	0.2–0.5 mL	Positive inotropic agent.
Furosemide	SC, IM, IV	1–4 mg/kg	Diuretic.

SC, subcutaneous; IM, intramuscular; IV, intravenous

euthanasia

Euthanasia is defined as a method of killing an animal that ensures minimal physical and psychological suffering.[71] The method must be consistent with recommendations of the American Veterinary Medical Association.[71] Barbiturate overdose is a commonly used and preferred method for euthanizing ferrets. These agents (e.g., pentobarbital and phenytoin combinations such as Euthasol) can be given intravenously (when an IV catheter is placed) or intraperitoneally. The intracardiac route may be used in animals under deep anesthesia. Carbon dioxide (CO_2) may be used to euthanize small ferrets. High concentrations of CO_2 and longer exposure times may be necessary for neonatal ferrets, as they are more tolerant of higher CO_2 concentrations than older ferrets. A recent report indicates that pre-filling the euthanasia chamber causes less stress in ferrets.[72] Secondary methods of ensuring euthanasia include inducing a bilateral pneumothorax with a blade after barbiturate overdose.

Table 19: Other Anesthetic Agents in Ferrets[70]

Drug	Route	Dosage	Comments
Sodium pentobarbital	IV, IP	30–45 mg/kg	Traditionally administered, IP can be used for terminal studies. May be used by IV route. Surgical plane difficult to control when administered IP. For long-term anesthesia, supplemental doses may be administered either IP or IV. Pentobarbital is not an analgesic drug.
Sodium thiopental	IV	8–12 mg/kg (with premedication)	Used for anesthetic induction by IV route. Depresses cardiac contractility and causes respiratory depression or even apnea. Thiopental is not an analgesic drug. Use 2% concentration solution.
Propofol	IV	6–8 mg/kg 1–3 mg/kg (with premedication)	Used for anesthetic induction by IV route. Depresses cardiac contractility; causes respiratory depression and possibly apnea. Causes less irritation than thiopental in perivascular injection.
Urethane	IP, IV	1.5 mg/kg	Long-acting hypnotic agent. Carcinogenic. Consider the use of analgesic for invasive procedures. Acceptable only for terminal studies.

IV, intravenous; IP, intraperitoneal

General criteria for euthanasia of moribund animals include:

- Rapid weight loss in a short time (often >20%)
- Extended period of weight loss progressing to emaciation
- Significant blood loss or persistent anemia leading to debilitation
- Evidence of irreversible organ failure
- Prolonged, intractable diarrhea or vomiting that is unresponsive to treatment
- Persistent difficult, labored breathing or respiratory distress
- Central nervous system clinical signs such as convulsions, head tilt, circling paralysis, and paresis
- Clinical signs of suspected infectious disease requiring a necropsy for diagnosis
- Significant and sustained decrease of body temperature
- Inability to reach food and water

experimental methodology

restraint

Proper restraint of a laboratory ferret is critical in order to ensure the safety of both the animal and the technical staff performing the procedures. The appropriate form of restraint (manual, mechanical, or chemical) depends on the nature of the procedure being performed and the disposition of the ferret.

Manual restraint is most commonly used when working with ferrets. It is strongly recommended that when working with alert laboratory ferrets, the handler don the appropriate personal protective equipment in order to avoid injury from bites and scratches. Frequent handling may reduce the incidence of aggressive behavior; however, under laboratory conditions behavior can be unpredictable, and the use of bite protection is highly advised. A variety of bite and/or scratch resistant gloves are available that afford protection while allowing sufficient dexterity to safely handle the animals. Ferrets can be removed from the cage by gently "scooping" the animals up just behind the front limbs and with the ribs supported by the palm of the hand. The animal can then be "scruffed" with the opposite hand by gently grasping the loose skin behind the head and above the shoulders on the dorsal side (**Figure 14**). Once a firm grip has been established, the hand supporting the ventral side of the animal can be removed and used to support the hind limbs while vertically holding the animal in an upright position. Proper scruffing technique results in the animal assuming a relaxed or limp, docile demeanor often accompanied by a "yawning" response (**Figure 15**). Further

Fig. 14 "Scruffing" a ferret.

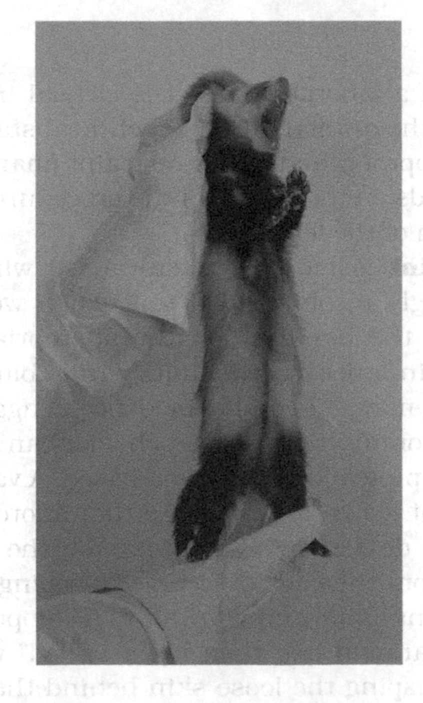

Fig. 15 "Yawning" response in a relaxed ferret.

restraint can be accomplished by maintaining the scruff of the animal and positioning the animal ventrally while cradling the body between the handler's arm and chest. This position is useful when administering liquids via the oral or nasal routes. While the animal will remain docile while being restrained by the scruff, care must be taken to partially support the lower extremities to avoid the potential occlusion of the airway due to the weight of the animal exerting pressure against the trachea when the animal is dangled for more than a few moments (**Figure 16**). Common procedures involving manual restraint include physical examination and/or clinical observation, oral gavage, and intramuscular or subcutaneous injections. Under these conditions, it is recommended to have one technician perform restraint while a second individual performs the procedure. This will allow the restrainer to employ the use of the appropriate protective equipment, while the person performing the procedures may do so with increased dexterity and safety.

Mechanical restraint is recommended for procedures requiring more than a few minutes to accomplish. There are a number of devices and techniques for mechanical restraint, which can be as simple as wrapping the animal in a towel while allowing the head to be exposed, to more complex devices such as restraint tubes or cones. Depending on the laboratory conditions, the number of animals to be restrained, and the need to effectively decontaminate equipment, any of these methods can be utilized. "Swaddling" the ferret by snuggly wrapping it in a towel with its forelimbs pressed against its body

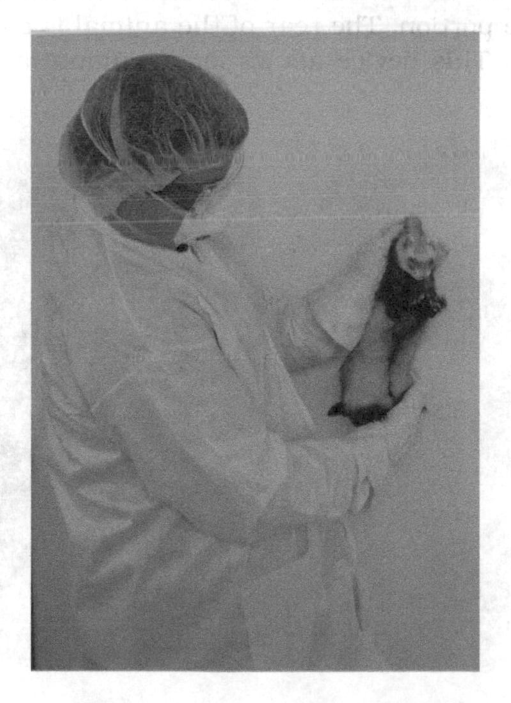

Fig. 16 "Scruffing" with proper support of lower extremities.

will restrain the animal sufficiently to perform some procedures. A restraint tube allows for a more secure immobilization of a ferret by inserting the animal head-first into a clear Plexiglas tube ending in a conical nosepiece with an aperture that allows for adequate breathing and a "butt plate" that can be positioned to secure the rear of the animal, thus preventing it from backing out of the tube (**Figures 17** and **18**). It is recommended that these tubes be clear (transparent), in order to allow technicians to observe the animals' breathing and behavior during restraint. Animals exhibiting signs of distress should immediately be removed from the restraint device. Ferrets are natural burrowers, and will generally enter these tubes willingly. Depending on the duration of time animals will need to reside in the restraint tube (generally not to exceed 30–45 minutes), it is recommended that the animals be conditioned to the tubes gradually over several days by increasing the amount of time spent in the tubes by 5–10 minutes per day. Another mechanical restraint device is the tamarack cone (**Figures 19–21**), which consists of a gradually narrowing collapsible cage near the front of the cone, with a canvas (or equivalent) zippered bag on the rear of the device. Similar to the restraint tube, the animal is introduced nose first into the cone portion of the device and is maneuvered until the head is toward the end of the cage portion. The rear of the animal is contained within the bag region. This device allows for maximum airflow and heat

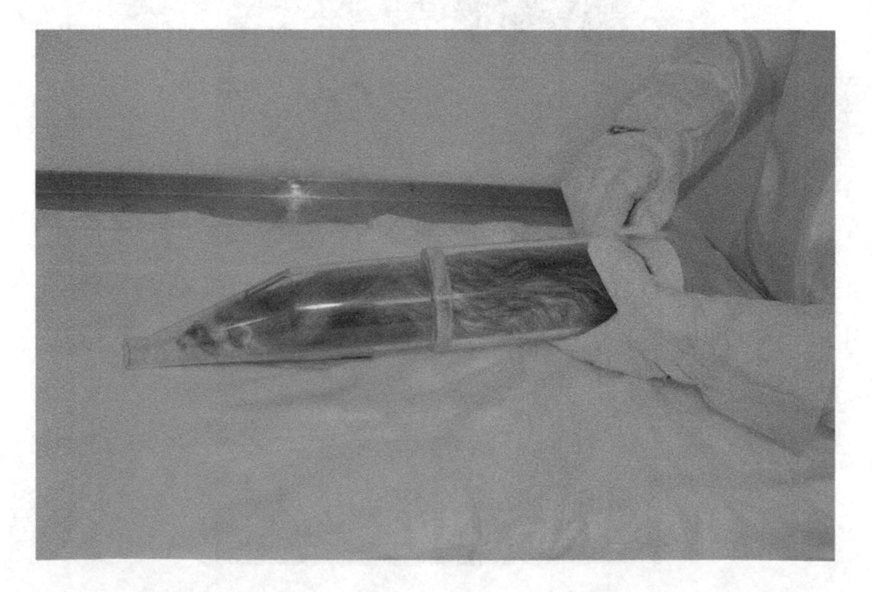

Fig. 17 Ferret in restraint tube with "butt plate" in position.

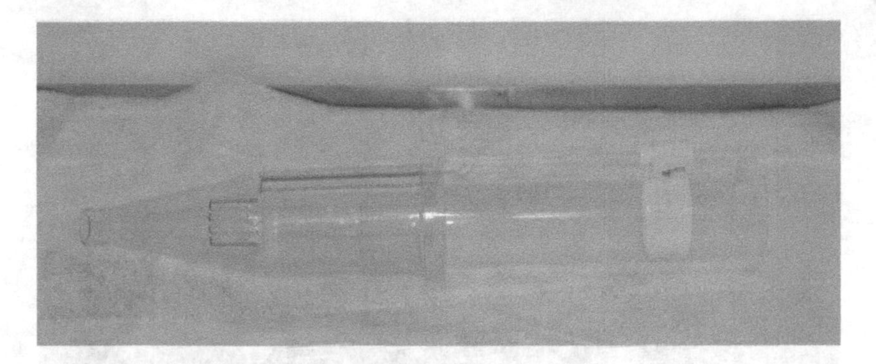

Fig. 18 Restraint tube with white "butt" plate.

Fig. 19 Tamarack cone.

diffusion while providing access to the cranial half of the animal for sample administration, injection, etc. The animal is quickly released by unzipping the bag portion of the device and carefully backing the animal out of the cone portion.

Chemical restraint is often employed for procedures requiring the animal to be completely immobilized in order to conduct sample collection or administration safely. Examples of this are intranasal and intratracheal administration and cranial vena cava blood collection. Drugs used for chemical restraint are discussed in Chapter 4, under the section Anesthesia. Injectable anesthetics generally require a longer recovery time (30–60 minutes), while inhaled anesthesia is rapidly reversible, with animals recovering in 3–5 minutes. Typical drugs used for chemical restraint include injectable anesthetics such as ketamine, ketamine/xylazine cocktail, and telazol, and inhaled anesthetics such as isoflurane. As with all anesthetics, adequate monitoring is critical to ensure animals are maintaining adequate

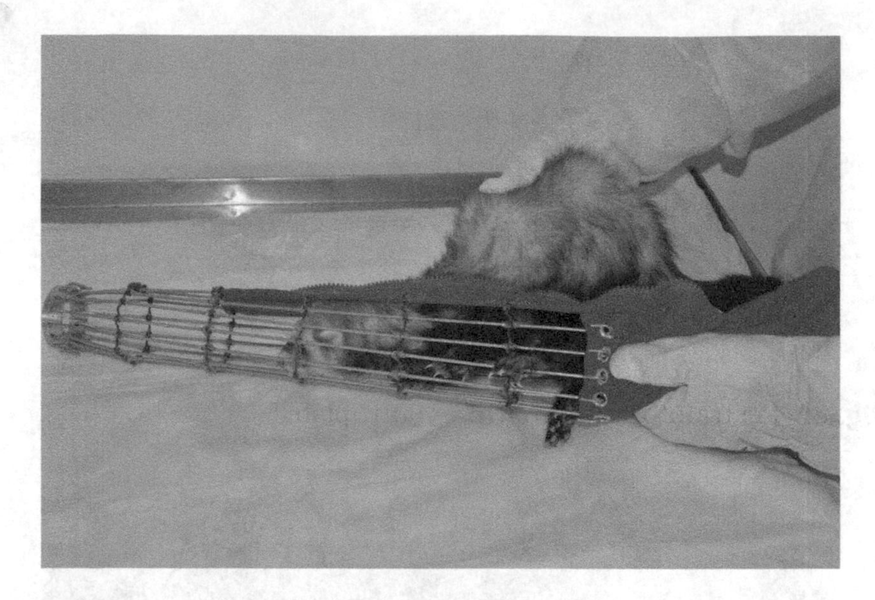

Fig. 20 Loading ferret into tamarack cone.

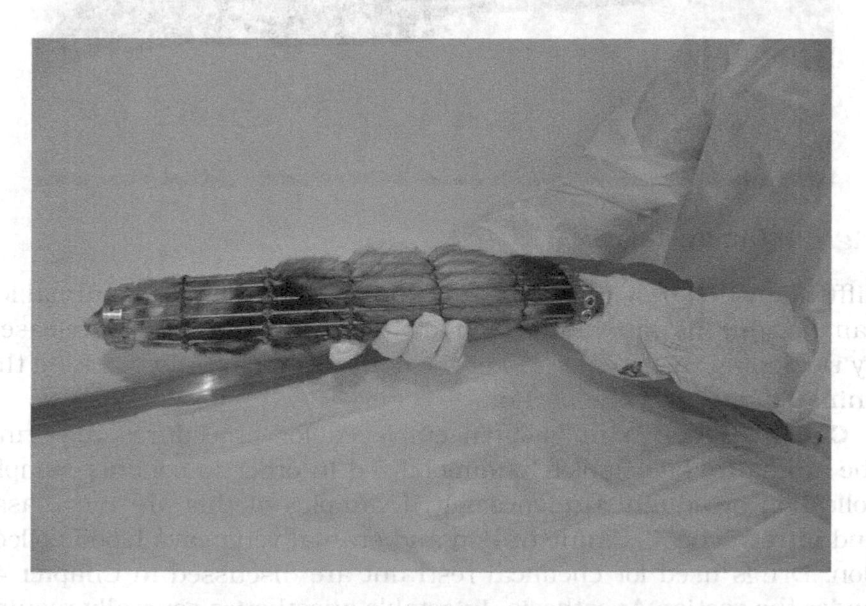

Fig. 21 Proper positioning of ferret in tamarack cone.

body temperature and respiratory function. The use of some anesthetics can result in transient decreases in body temperature (telazol and ketamine) or affect hematological parameters (isoflurane). When using injectable anesthetics that decrease body temperature, animals should be allowed to recover on non-metallic surfaces such as a soft towel, as metal surfaces can leach body heat from the animals. If isoflurane is being used to collect blood samples for clinical pathology, it should be noted that isoflurane anesthesia has been reported to result in transient reduction in hematocrit, red blood cell (RBC) count, and hemoglobin concentration within 15 minutes, with theses values returning to normal within 45 minutes.[92] These changes result from temporary sequestration of RBCs within the spleen.[93]

sampling techniques

Blood Collection

The method and site of blood collection depend on several variables, including the skill of the technician, the volume of blood to be collected, and the frequency of blood collection. Smaller volumes of blood (<1.0 mL) can be collected from the saphenous or cephalic veins, or even via the retro-orbital plexus. Larger blood volumes (>1.0 mL) are generally collected via venipuncture of the jugular vein or the cranial vena cava. Under some circumstances, such as pharmacokinetic studies, multiple blood collections must occur in a short period of time without the use of anesthesia. Under these circumstances, it is recommended to consider the use of surgically implanted indwelling catheters, such as those equipped with venous access ports.[94-96] The blood volume of a ferret is approximately 5–6% of the total body weight, and no more than 10% of the total blood volume should be collected in a single blood collection. Within a two-week period, no more than 20% of the total blood volume should be collected.[97-99]

Cranial vena cava (CVC) blood collection is the preferred method for volumes greater than 1 mL of blood. While this technique can be performed on ferrets that are manually restrained, it is recommended to perform this procedure on chemically restrained ferrets for maximum safety of the ferret and the handler. The CVC venipuncture involves a "blind stick," requiring a thorough understanding of the vasculature and associated landmarks for proper needle placement (**Figure 22**). The anesthetized ferret is placed in dorsal recumbency, and the site is swabbed with alcohol in order to maintain proper aseptic technique. For increased visualization, the chest should be shaved with clippers and aseptically prepared with an iodine or chlorhexidine scrub followed by isopropyl alcohol. A 25-gauge needle attached to a 5–10-mL syringe is carefully placed just lateral to the manubrium and cranial to the first rib, slowly

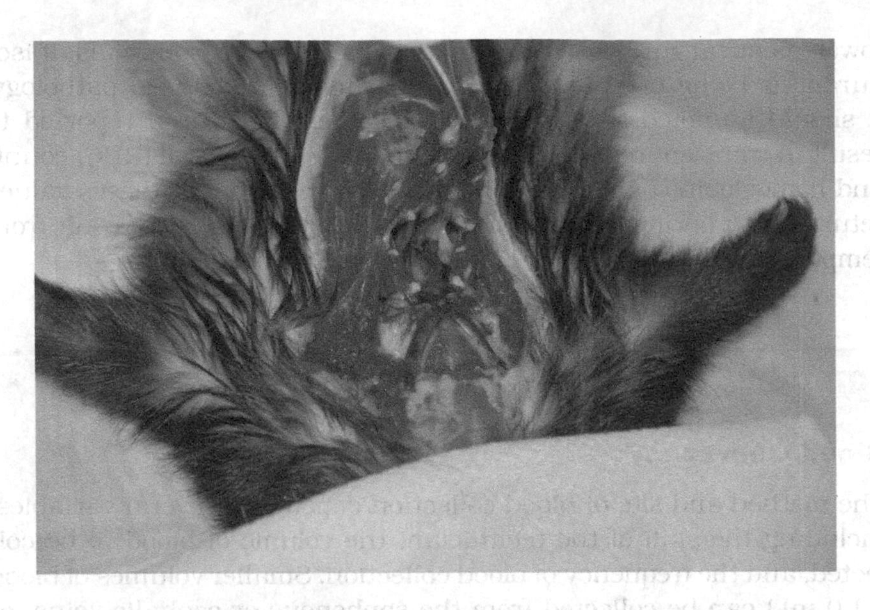

Fig. 22 Underlying vascular anatomy for blood collection.

inserted at approximately a 45-degree angle, and directed toward the opposite hind limb[100] (approximately ½ to 1″ deep) (**Figure 23**). Gentle negative pressure is applied on the syringe until a "flash" of blood is visible in the needle hub (**Figure 24**). Blood is then slowly withdrawn by gently pulling back on the plunger while maintaining the position of the needle until the required blood volume is collected (**Figure 25**). Once the volume is collected, the needle is gently removed, and pressure is applied to the injection site for approximately 30 seconds to prevent the formation of hematomas. Large volumes up to complete exsanguination can be performed using this method.

Jugular venipuncture can be performed on manually, mechanically, or chemically restrained ferrets, and is typically employed for blood collection volumes of greater than 1 mL. The animal is placed in dorsal recumbency, as in CVC venipuncture. The external jugular veins are best visualized when the region from the sternum to just below the neck is shaved. The jugular veins are fairly superficial and extend craniolaterally from just below the base of the ears to the midline at the thoracic inlet. The vein can be palpated by using the index finger and thumb to occlude the vein above and below the point of needle insertion. The blood collection site should be prepared for aseptic collection as previously described. Once the vein is distended, a 20–23-gauge needle attached to a 5–10-mL syringe is inserted at approximately a 30-degree angle by first

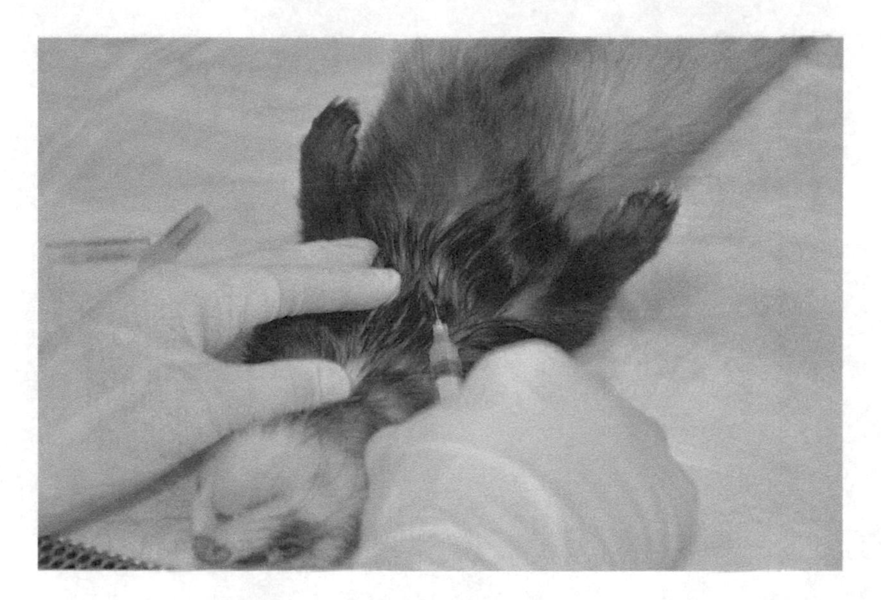

Fig. 23 Proper placement and angle of needle for blood collection.

piercing the skin, and then directing toward the vessel.[101] Once the vessel is pierced, blood is slowly collected, and the needle is removed as described above for CVC blood collection.

Cephalic venipuncture can be approached in much the same way as in rabbits, dogs, and cats, and is used for collection of <1.0 mL of blood. This method requires the help of an additional technician who manually restrains the animal. It should be noted that this vessel is small and superficial and ferret skin is quite tough, and so this method is considered moderately difficult. Visualization of the vessel may be improved by clipping the site of hair. A tourniquet (surgical tubing and alligator clip) is applied above the elbow to allow the cephalic vein to become distended. The site should be prepared with an alcohol wipe prior to injection to ensure proper aseptic technique. Once palpated, a 25–27-gauge needle attached to a 1-mL syringe is injected through the skin and into the cephalic vein. Blood (<1.0 mL) is slowly collected by applying slight negative pressure through withdrawal of the syringe plunger while holding the leg steady and simultaneously removing the tourniquet with the other hand.[101-103]

Saphenous venipuncture for collection volumes of <1.0 mL is performed similar to cephalic vein collection. The animal is placed in either ventral or dorsal recumbency, and the rear leg is restrained just distal to the stifle to distend the vessel using the thumb or fingers. A 26-gauge needle attached to a 1.0-mL syringe is inserted into the

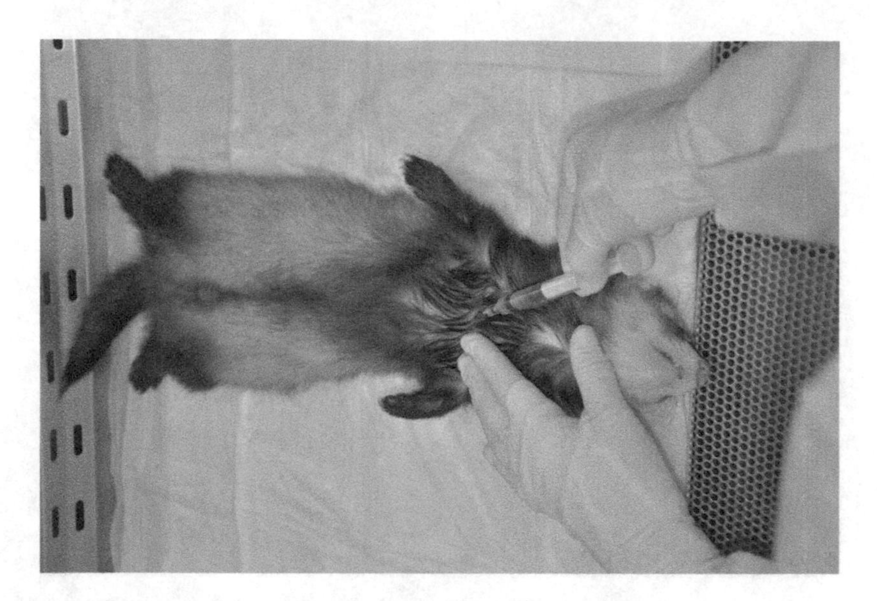

Fig. 24 Sample collection in progress.

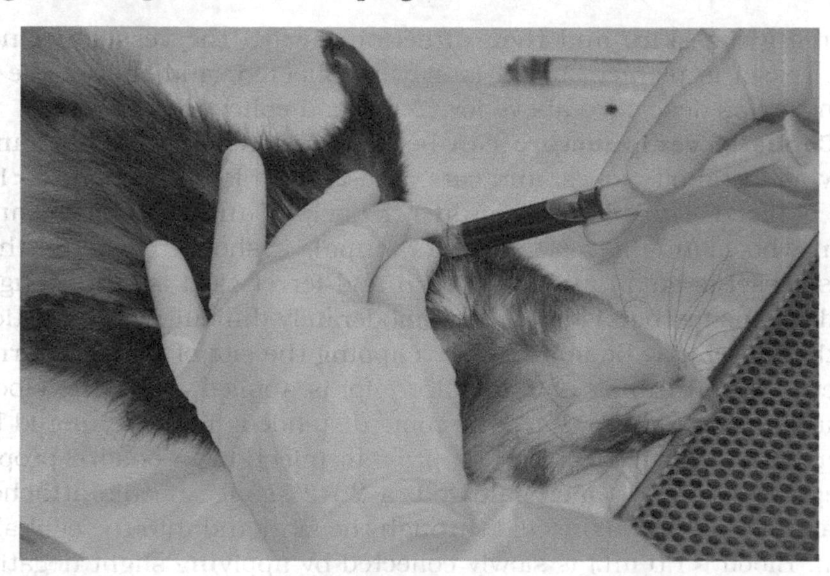

Fig. 25 Completed blood collection.

vessel, taking care not to roll the vein. Blood is slowly collected, and the needle is removed and the site held off, as previously described.

Retro-orbital blood collection has been reported as a means of collecting volumes of 1–10 mL. This method should be performed only on anesthetized ferrets by highly skilled operators. The anesthetized ferret is held with its head angled downward, being supported by the technician's hand and forearm. The animal's eye is partially protruded by pulling the eyelid open while applying pressure with the first two fingers periorbitally. A heparinized capillary tube that has previously been broken is placed with the sharp end in the medial canthus. The tube is gently rotated and pressed through the fat and fibrous tissue into the retro-orbital plexus with gentle yet firm pressure. Blood then flows from the capillary tube into an appropriate container. Other blood collection sites are preferable to the retroorbital approach and require less operator expertise. As a result this technique is seldom used.

Cardiac puncture can be used to collect large volumes of blood, and should be performed only by experienced personnel on deeply anesthetized ferrets. This procedure, if performed incorrectly, is very dangerous, and should be performed on ferrets only directly prior to euthanasia. It should be noted that large blood volumes can also be collected via CVC instead of cardiac puncture, up to and including complete exsanguination.

Washes and Swabs

Nasal washes and pharyngeal swabs are often collected to measure virus shedding in infected animals. The nasal wash can be collected on alert or anesthetized animals; however, performing this procedure under anesthesia is preferred for the safety of the animal.

Nasal wash fluid is typically composed of a normal saline solution supplemented with antibiotics (gentamicin, penicillin, and/or streptamicin) and in some cases a protein source to increase the stability of collected virus (bovine serum albumin). A syringe is fitted with a 22-G catheter from which the needle has been removed. Approximately 2 mL of nasal wash solution is drawn into the catheter. While scruffing the ferret and holding the head over the collection cup (a urinalysis cup works well), the catheter is placed into the groove of the naris and ½ of the solution is gently flushed into the nasal cavity (**Figure 26**). The second ½ of the fluid is flushed into the other naris. The positioning of the ferret's head and gravity allow the wash to flow into the collection cup. Samples can be aspirated from the collection cup and transferred to a storage vial using the same cathetered syringe.

Pharyngeal swabs are collected from alert ferrets by first placing into the mouth an oral speculum that is fitted with an opening for insertion of the swab. The procedure requires two technicians. One technician manually restrains while a second technician places the

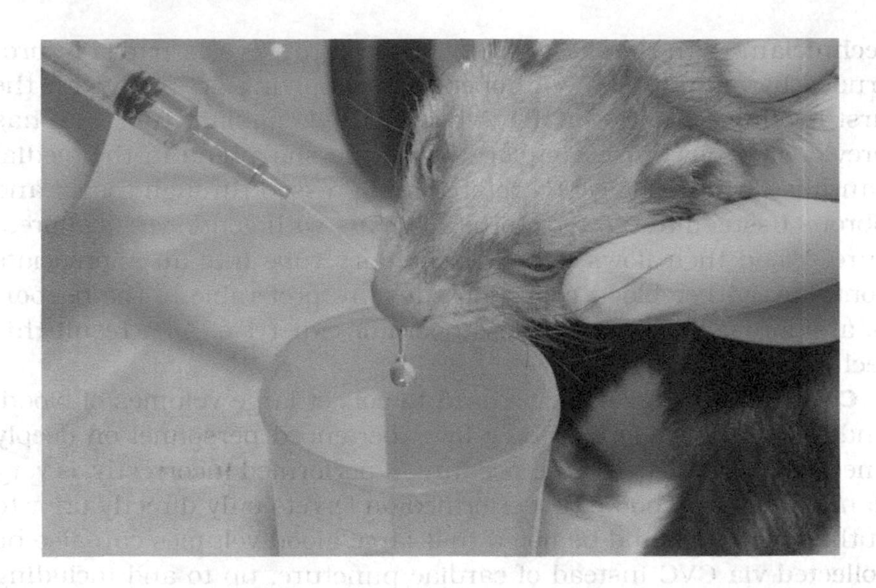

Fig. 26 Collection of nasal wash.

oral speculum in place and inserts the swab through the opening and into the pharyngeal region. A rapid swabbing of the pharyngeal region (1–2 seconds) is sufficient to collect the material (**Figures 27** and **28**), which is typically placed into a 1-mL tube containing the same solution as the nasal wash.

Urine Collection

There are several means of urine collection in the ferret. Ferrets may be placed into metabolism cages for easy collection of urine. This is the least invasive method of urine collection, but it does not yield a sterile sample for analysis. Cystocentesis and catheterization collections will provide a sterile sample for analysis, and aseptic techniques and procedures should be used to collect the sample.

Cystocentesis may be performed if the urinary bladder is palpable. Restraint (manual or chemical) is critical during this procedure, and when performed on alert animals, distraction with a sweet substance such as Nutrical has been successful. Using a 3-mL syringe and a 1-inch, 22- or 23-gauge needle, the bladder is accessed through the lateral body wall,[102] and urine is collected by gentle aspiration via syringe. Isoflurane anesthesia is preferred for this procedure when performed on a healthy ferret.

Urinary bladder catheterization has been reported[7,104] as a successful means of urine collection in both male and female ferrets.

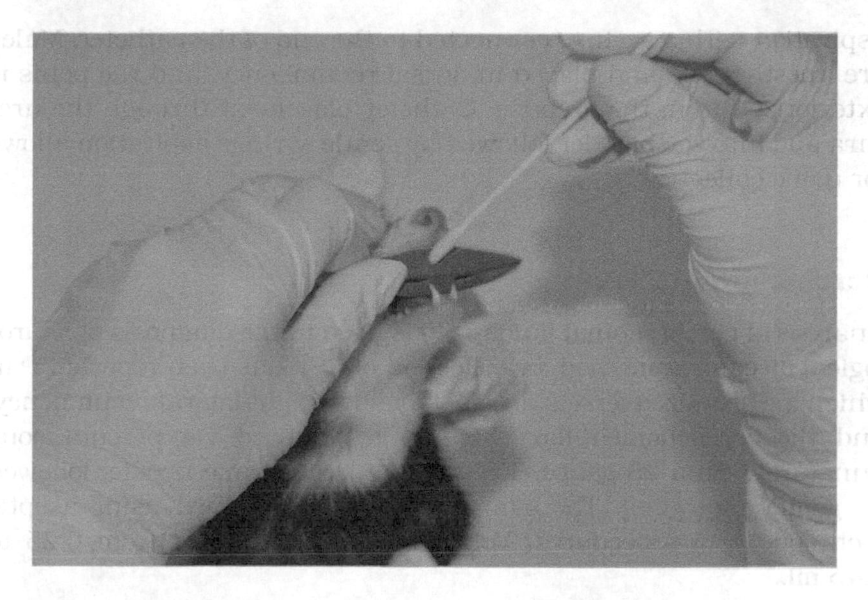

Fig. 27 Oral speculum in place for a pharyngeal swab.

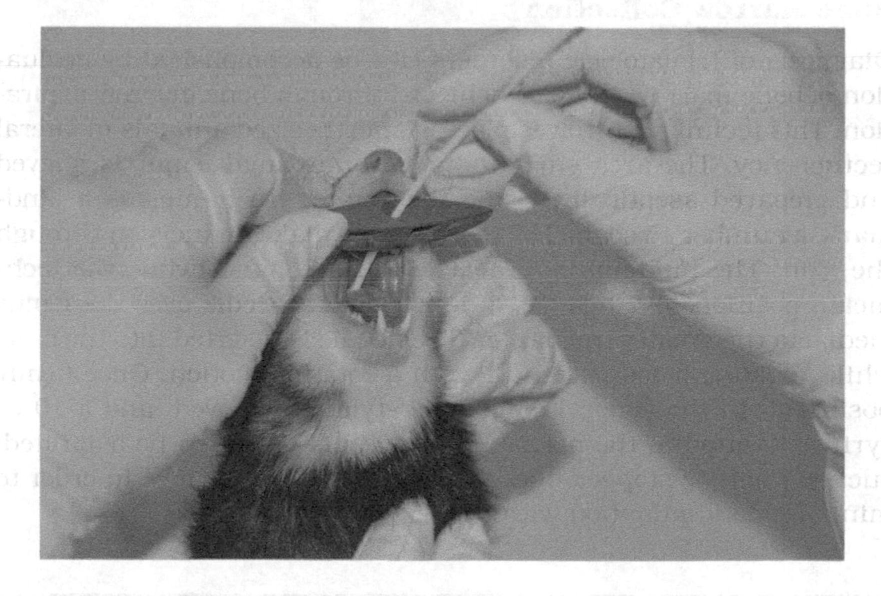

Fig. 28 Collecting pharyngeal swap.

The procedure is performed under sedation (chemical restraint) using a 3.5 French rubber urethral catheter. Females are positioned in ventral recumbency with the pelvis slightly elevated with a rolled towel. With the aid of a vaginal speculum, the catheter is advanced into the urethra to the bladder, and urine is collected using gentle aspiration with a syringe connected to the end of the catheter. Males are anesthetized and placed in dorsal recumbency, and the penis is exteriorized from the prepuce. Catheter placement through the urethra and into the bladder followed by gentle syringe aspiration allows for urine collection.

Cerebrospinal Fluid

Analysis of cerebrospinal fluid (CSF) is used in the diagnosis of neurological disease. A method for collection of CSF has been reported[105] in which anesthetized ferrets are positioned in right lateral recumbency, and the cerebellomedullary cistern is accessed via percutaneous puncture with a 25-gauge, 1.6-cm-long hypodermic needle, followed by gentle aspiration. The sample should be obtained using aseptic technique and procedures. Volumes collected ranged from 0.25 to 0.45 mL.

Bone Marrow Collection

Diagnosis of hematologic disorders can be accomplished by evaluation of bone marrow samples achieved through bone marrow aspiration. This technique involves placing anesthesized animals in lateral recumbency. The area surrounding the proximal femur is shaved and prepared aseptically. Using the greater trochanter as a landmark, a number 15 scalpel blade is used to make an incision through the skin. The hindlimb is grasped with one hand, while the technician positions a 20-gauge, 1.5-inch spinal needle over the femur medial to the greater trochanter. The needle is inserted into the bone while applying steady pressure with a rotating motion. Once firmly positioned in the femoral canal, the stylet is removed, and a 10-cc syringe is fitted to the needle. Marrow fluid can then be aspirated; suction must be stopped once marrow enters the syringe in order to minimize contamination with blood.

administration techniques

Delivery of compounds can be achieved through various routes, depending on the volume, pH, solubility, and intended dosing route. Typical sample volumes for a 1-kg ferret are listed in **Table 20**.

Table 20: Recommended Volumes and Needle Gauges for Sample Administration[a]

Per os (oral) mL		Subcutaneous (mL)[b]		Intramuscular (mL/site)[b]		Intraperitoneal (mL)[b]		Intravenous (mL)[c]	
recommended	maximum	recommended	maximum	recommended	maximum	recommended	maximum	bolus	slow injection
5	15	10	20–30	0.25	0.5	5	20	2–5	10

[a] adapted from Cornell University Center for Animal Resources and Education
[b] ≤21G needle
[c] ≤23G needle

Subcutaneous (SQ, SC)

The area between the shoulder blades should be shaved and prepared aseptically with iodine or chlorhexidine scrub and isopropyl alcohol. The skin is tented, and the needle is inserted through the skin into the subcuticular space above the underlying muscle tissue (**Figures 29** and **30**). The syringe is gently aspirated, and if no blood enters the syringe, the material is administered subcutaneously.

Note: Care should be taken to avoid injection into the SQ fat pad, which may slow absorption of some compounds.

Intraperitoneal (IP)

Injections into the peritoneum are performed by inserting the needle into the animal's lower right quadrant so as to avoid injury to vital organs (**Figure 31**). Typically, a 23-gauge, 1.5-inch needle is placed lateral and posterior to the umbilicus. The ferret may be slightly inverted (head down) in order to facilitate proper delivery. The gauge of the needle will depend on the viscosity of the compound being delivered. The needle should enter the abdomen at a shallow angle, and once positioned into the peritoneum, the technician should gently aspirate. If no aspirate enters the needle, the compound can be slowly administered into the peritoneum. If aspiration of intestinal contents occurs, contamination has occurred and the needle and syringe should be removed and disposed of.

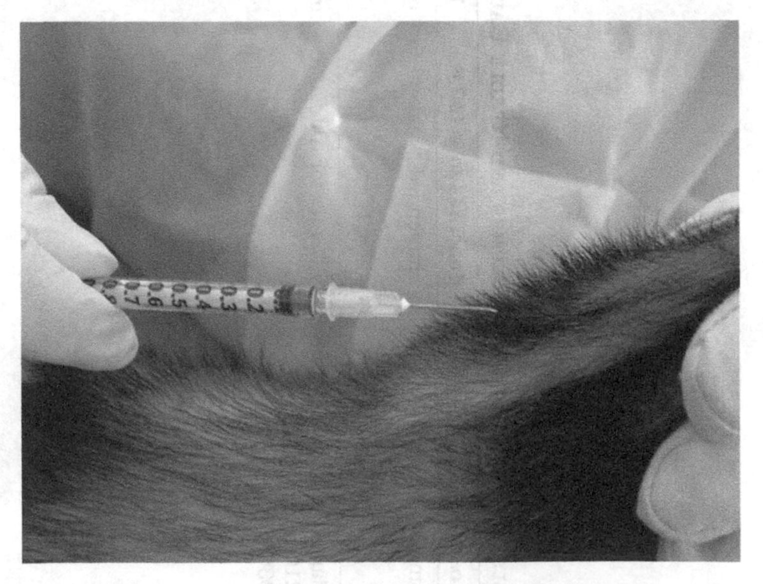

Fig. 29 Subcutaneous injection between shoulder blades.

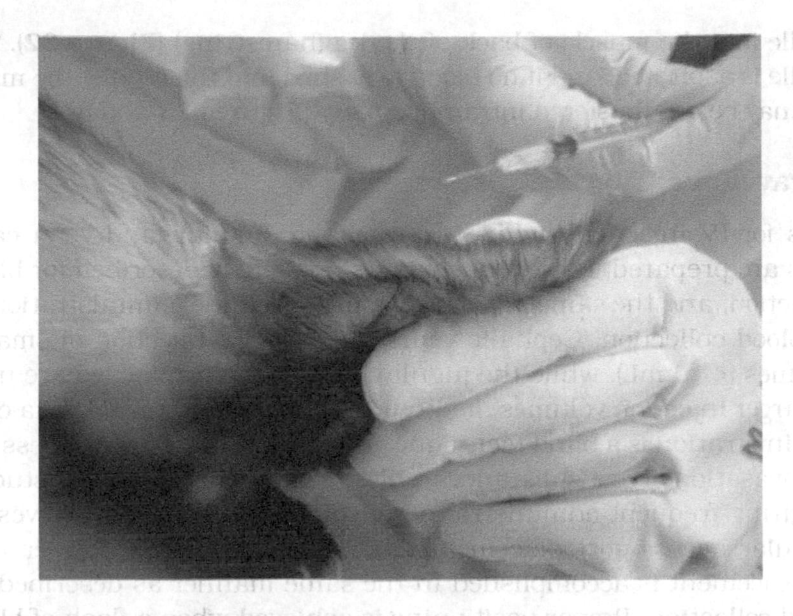

Fig. 30 Subcutaneous injection.

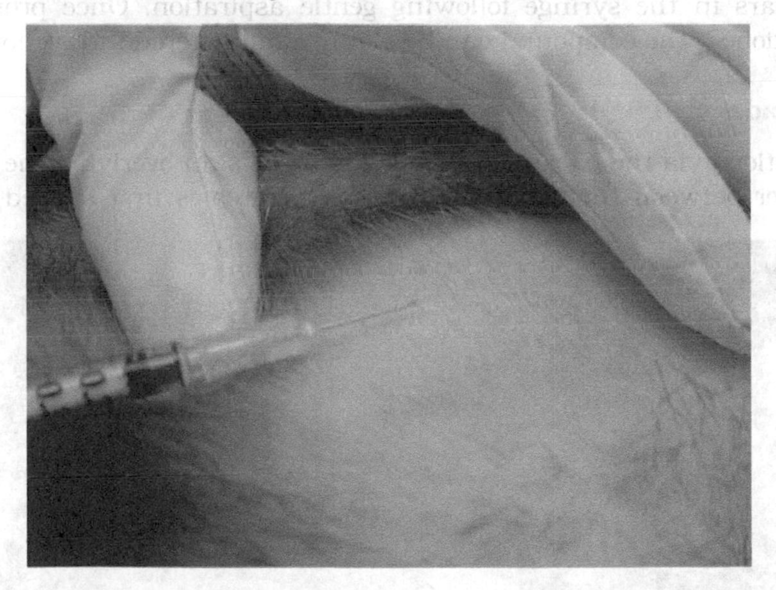

Fig. 31 Interperitoneal injection in lower abdomen.

Intramuscular (IM)

The hindlimb muscles are common sites for IM administration in ferrets. Needle placement should not be too deep, in order to avoid the femoral nerve and femur. The ferret should be restrained by one operator, while the second operator extends the hindlimb of the animal and places the needle into the muscle of back of the leg (hamstring) (**Figure 32**). The needle is aspirated, and if no blood is flashed into the needle, the material may be administered intramuscularly (**Figures 33** and **34**).

Intravenous (IV)

Sites for IV injection include the cephalic, jugular, and vena cava. Sites are prepared in the same way as previously described for blood collection, and the same principles apply to volume administration as for blood collection. Cephalic veins are used for injection of smaller volumes (<1.0 mL), while the jugular and cranial vena cava are used for larger injection volumes. It should be noted that cranial vena cava administration is a blind technique and it is not possible to assess the extravasation of test substances when given by this route. For studies requiring frequent administration of compound into a larger vessel, vascular access-port placement should be considered. Proper needle placement is accomplished in the same manner as described for blood collection. Proper positioning is achieved when a flash of blood appears in the syringe following gentle aspiration. Once properly positioned, the compound can be slowly administered intravenously.

Intradermal (ID)

Injections via the ID route are given into the skin overlying the thorax or between the shoulder blades. The area is first shaved and

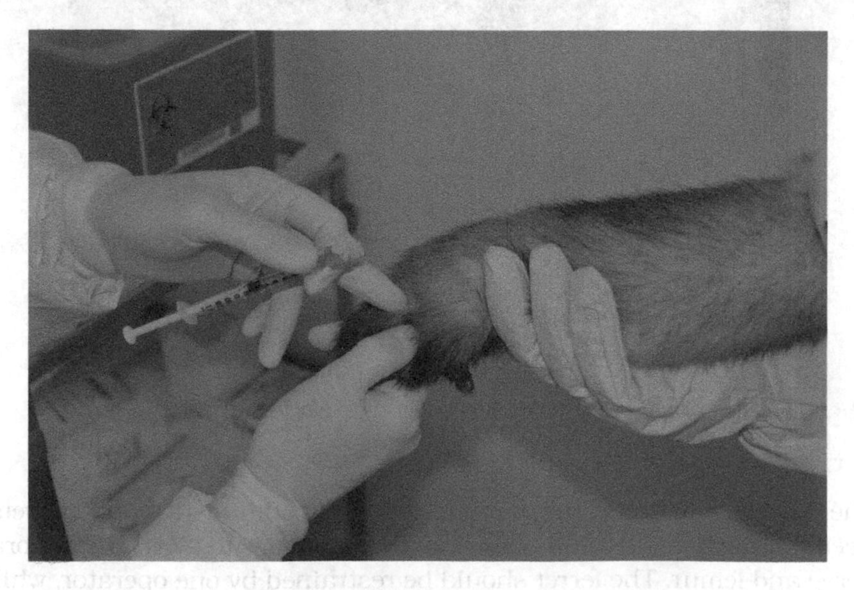

Fig. 32 Location for intramuscular injection.

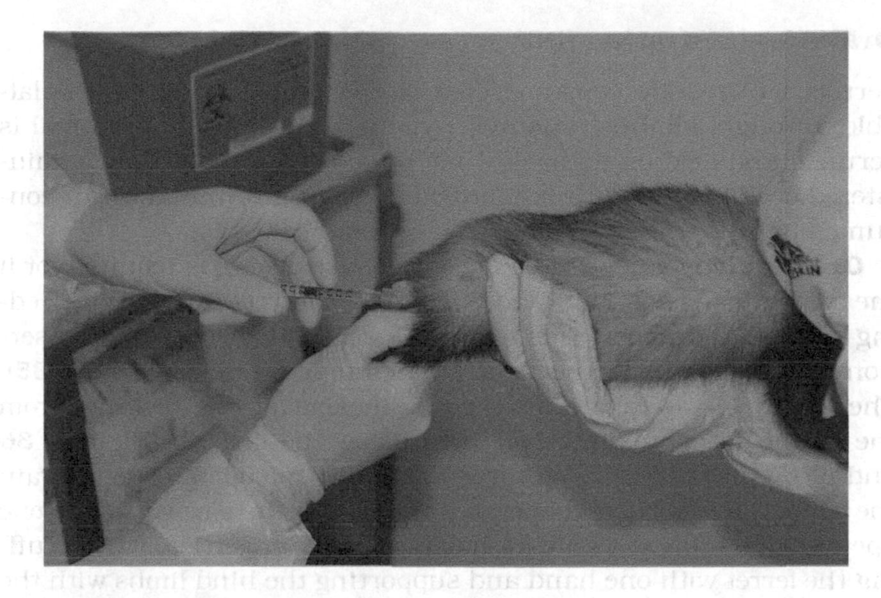

Fig. 33 Checking needle placement.

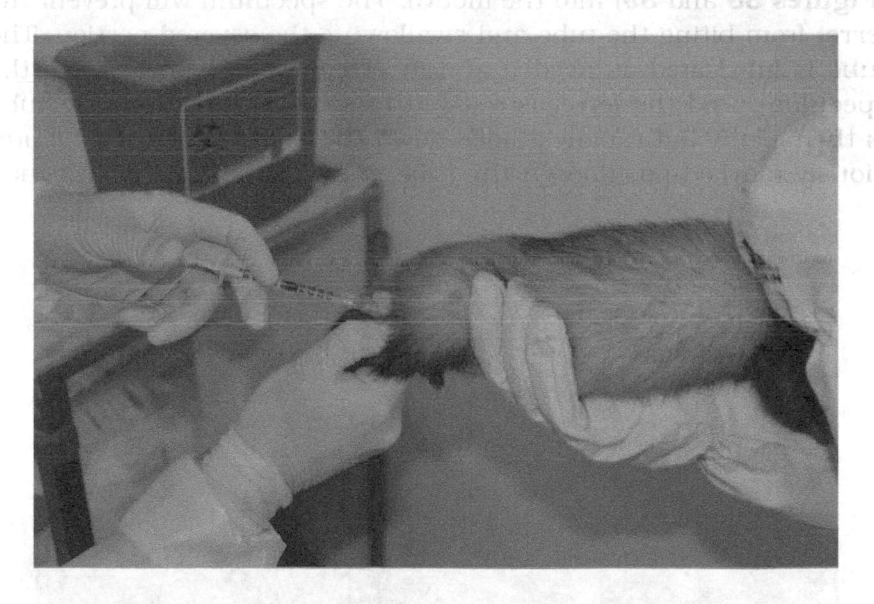

Fig. 34 Administering material via intramuscular injection.

prepped aseptically. A 23-gauge or smaller needle is inserted at an acute angle such that only the bevel of the needle is inserted into the dermis, avoiding the subcutaneous tissue. The compound is then gently administered intradermally. Proper placement is confirmed by the formation of a small bleb at the injection site.

Oral Administration (PO)

Ferrets will readily consume compounds orally if they are palatable, through administration via syringe or dropper. The animal is scruffed and held on its back while the compound is slowly administered orally. Adults drink approximately 10 mL/min and can consume up to 100 mL during a single session.[106]

Orogastric gavage can be used if precise dosing is required, or if the compound is not readily accepted by the ferrets through the feeding syringe or dropper. This method is accomplished through insertion of a catheter no farther than 5 mm into the stomach (**Figure 35**). The length of the tube is marked by measuring the distance from the mouth to approximately 1 inch below the last rib (**Figures 36** and **37**). This procedure requires the use of one operator to restrain the animal and a second operator to perform the gavage. While one operator restrains the animal, holding the ferret vertically by scruffing the ferret with one hand and supporting the hind limbs with the other, the second operator places an oral speculum, or "bite block," (**Figures 38** and **39**) into the mouth. The speculum will prevent the ferret from biting the tube and swallowing the severed portion. The tube is lubricated as needed and inserted through the hole in the speculum, and the ferret is allowed to swallow the tube. The tube is then gently but rapidly guided down the esophagus until the previously marked position on the tube is reached. The operator then

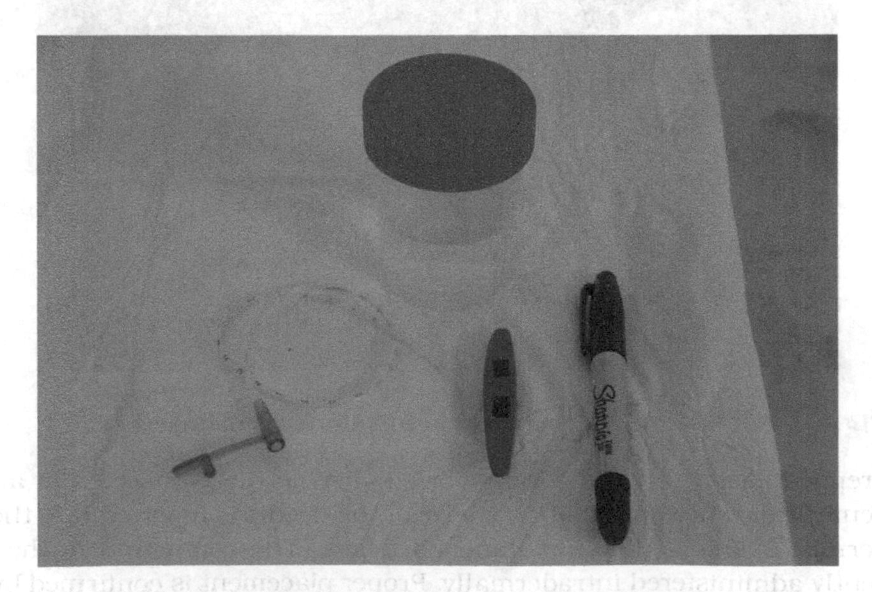

Fig. 35 Supplies for orogastric gavage.

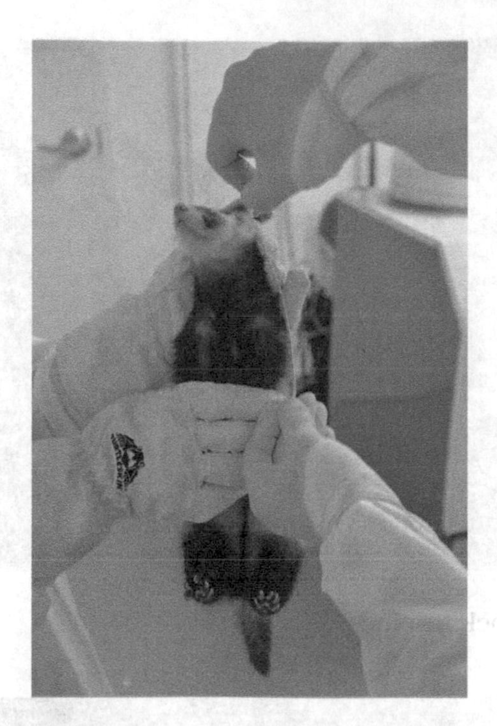

Fig. 36 Measuring for placement of catheter.

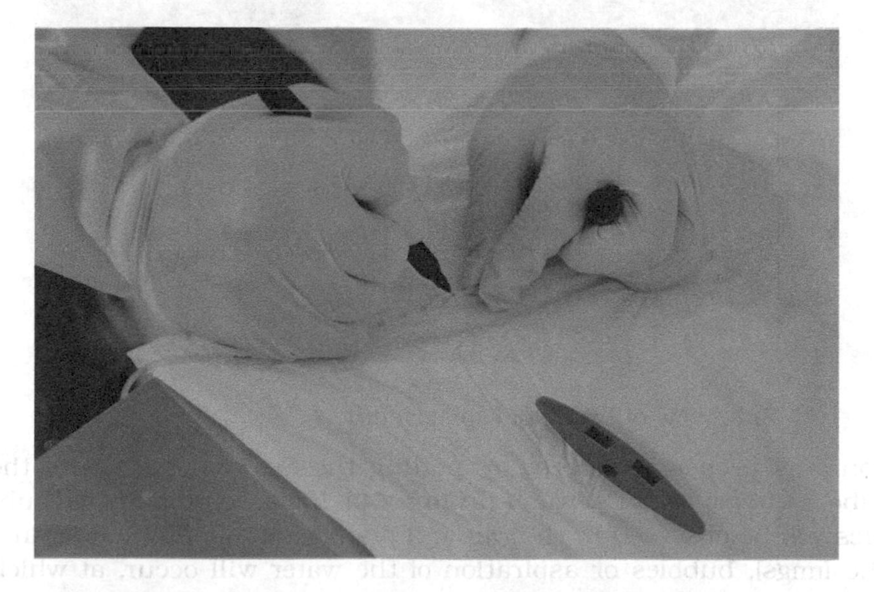

Fig. 37 Marking catheter prior to administration.

Fig. 38 Bite block in place.

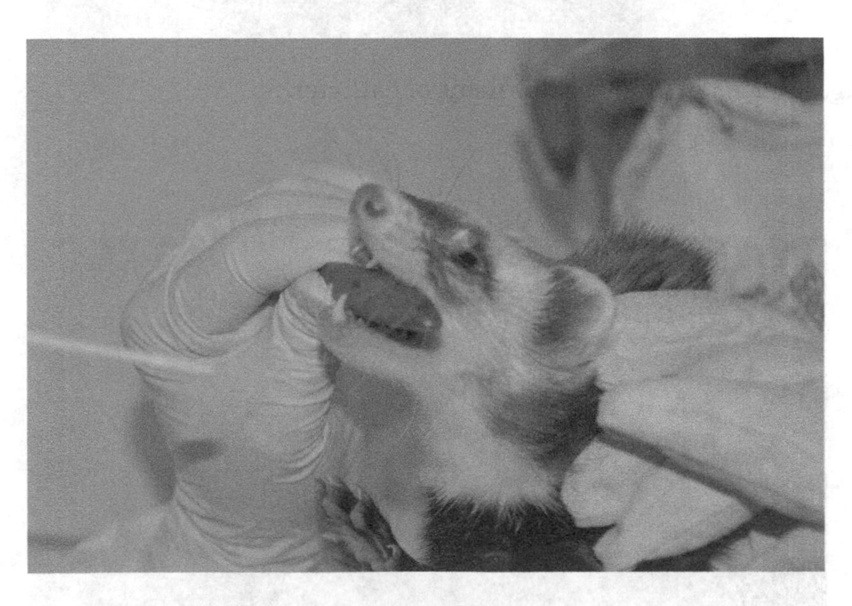

Fig. 39 Side view of bite block placement.

confirms proper placement by putting the external portion of the tube into a container of water (**Figure 40**). If placed properly, no bubbles will be observed in the water. If placed improperly (that is, into the lungs), bubbles or aspiration of the water will occur, at which point the tube is immediately removed. Once proper placement is confirmed, a syringe is attached to the tube, and the compound may be administered (**Figure 41**). Typical volumes are 5–15 mL and should never exceed 50 mL in a single dosing.[7]

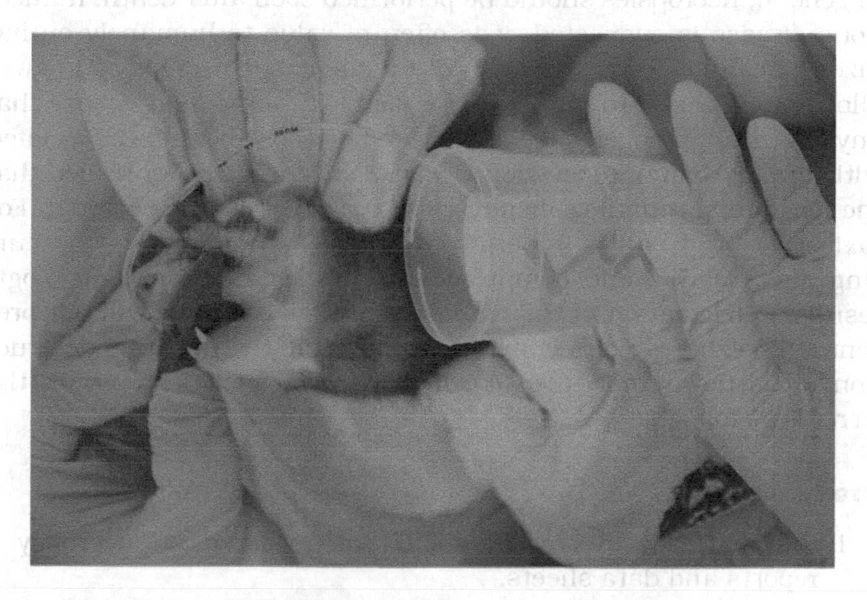

Fig. 40 Checking placement of catheter into the stomach.

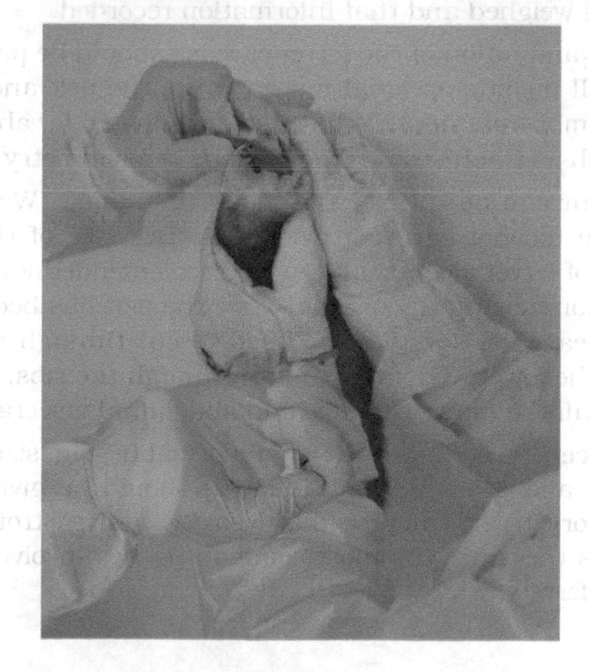

Fig. 41 Administering material.

necropsy

In general, necropsies should be performed soon after death. If infectious disease is suspected, it is often of value to humanely euthanize clinically ill ferrets and conduct a diagnostic necropsy. This will allow the prosector to take blood for serologic assay and ensure that any bacterial pathogens are not masked by those bacteria associated with postmortem degeneration (PMD). It is highly recommended that the laboratory animal veterinarian be consulted before necropsy. For toxicologic studies it is imperative to avoid PMD, as changes occurring after death could compromise the interpretation of histologic results. Refrigeration of the carcass will slow PMD but will not prevent it. Freezing is not recommended owing to the resulting destruction of tissue architecture. A brief description of necropsy in the ferret is as follows:

Procedure

1. Information and results should be recorded on necropsy reports and data sheets.
2. Prior to beginning the necropsy, the animal should be identified and weighed and that information recorded.
3. A full examination of the ferret carcass should be performed. This will include a visual inspection of the pelt and all orifices. Limbs and torso should also be palpated for abnormalities such as fractures or the presence of asymmetry.
4. Ferret fur can be thick and difficult to manage. Wetting the fur with alcohol can help with this. Because of the tough nature of ferret skin, scalpels are recommended for opening ferrets for viewing the organs. Once the pelt has been peeled back, heavy scissors can be used to cut through the muscles of the lower abdomen and up through the ribs. This will expose all the major thoracic and abdominal viscera.
5. The viscera are examined for abnormalities in size, shape, texture, and color. Special attention should be given to the respiratory tract, trachea, nasal cavity, and gastrointestinal tract, as these are the most common tissues involved in disease in ferrets.

6. Samples for routine histopathological interpretation can be immersed/perfused in an appropriate fixative (10% neutral buffered formalin, Bouin's solution, or 4% paraformaldahyde). The choice of the fixative will depend on the purpose of the necropsy. For example, a better choice of fixative for evaluation of the testes and eyes may be Bouin's solution. A ratio of one part tissue to 10 parts formalin should be used for routine fixation of tissues. Organs such as the lungs and intestinal tract should be inflated or the lumens infiltrated with formalin for adequate fixation. Regardless of the fixative, samples collected should be small, less then 1 cm thick if possible. This allows for adequate penetration of the fixative. The prosector should check with a pathologist or the researcher to evaluate the best method of preservation. Some tissues will be saved whole and may be larger than 1 cm. For example, the brain, skull, and bones, are sometimes sampled together as a unit. In such cases, a discussion with a pathologist about the best method for fixation is advised.

resources and additional information

Because this volume is intended to be a handbook, coverage is not exhaustive for most topics. In view of this, additional resources for information regarding the care and use of laboratory ferrets are provided here.

organizations

A number of professional organizations exist that can serve as initial contacts for obtaining information regarding specific issues about the care and use of laboratory ferrets. Membership in these organizations should be considered, since it allows the laboratory animal science professional to stay abreast of regulatory issues, improved procedures in the use of animals, management issues, and animal health issues. Relevant organizations include the following:

American Association for Laboratory Animal Science (AALAS), www.aalas.org, serves a diverse professional group, ranging from principal investigators and animal care technicians to veterinarians. The journals *Comparative Medicine* and *Journal of the American Association for Laboratory Animal Science* are both published by AALAS and serve to communicate relevant information. AALAS sponsors a program for certification of laboratory animal science professionals at three levels: assistant laboratory animal technician (ALAT), laboratory animal technician (LAT), and laboratory animal technologist (LATG). Further, a certification program for managers of animal resource programs has been developed. An extensive online resource, the AALAS Learning Library, offers subscribers an extensive menu of courses relevant to laboratory animal science. The association also sponsors an annual meeting and several electronic listserves, including TechLink for animal technicians, CompMed for professionals working in comparative medicine and biomedical research, and the IACUC Forum for IACUC members and staff. Local groups have also organized into smaller branches.

Laboratory Animal Management Association (LAMA) www.lama-online.org, serves as a mechanism for information exchange between individuals charged with management responsibilities for laboratory animal facilities. In this regard, the association publishes the *LAMA Review* and sponsors an annual meeting.

American Society for Laboratory Animal Practitioners (ASLAP), www.aslap.org, is an association of veterinarians engaged in laboratory animal medicine. The society publishes a newsletter to foster communication between members and sponsors sessions at the annual AALAS meeting and the annual meeting of the American Veterinary Medical Association.

American College of Laboratory Animal Medicine (ACLAM), www.aclam.org, is an association of laboratory animal veterinarians founded to encourage education, training, and research in laboratory animal medicine. The ACLAM board certifies veterinarians in laboratory animal medicine. The group sponsors the annual ACLAM Forum with its continuing-education meetings, along with sessions at the annual AALAS meeting.

Laboratory Animal Welfare Training Exchange (LAWTE), www.lawte.org, is an organization of people who train in and for the laboratory animal science field. By sharing ideas on methods and materials for training, members can learn together how best to meet the training and qualification requirements of national regulations and guidelines. LAWTE holds a conference every two years for trainers to exchange information on their training programs in the United States and abroad.

Institute for Laboratory Animal Research (ILAR), www.dels.nas.edu/ilar. The mission of ILAR is to evaluate and disseminate information on issues related to the scientific, technological, and ethical use of animals and related biological resources in research, testing, and education. Using the principles of refinement, reduction, and replacement (3Rs) as a foundation, ILAR promotes high-quality science through the humane care and use of animals and the implementation of alternatives. Through the reports of expert committees, the *ILAR Journal*, web-based resources, and other means of communication, ILAR functions as a component of the National Academies to provide independent, objective advice to the federal government, the international biomedical research community, and the public.

Association for Assessment and Accreditation of Laboratory Animal Care International (AAALAC International), www.aaalac.org. AAALAC is a nonprofit organization that provides a mechanism for peer evaluation of laboratory animal care and use programs. Accreditation by AAALAC is widely accepted as strong evidence of a quality research animal care and use program.

American Ferret Association (AFA), www.ferret.org, is a not-for-profit organization focusing on all aspects of ferret ownership. The website provides information on the care and showing of ferrets as well as veterinary and legislature regarding ferrets. The site also provides an extensive list of resources for ferret owners.

publications

A number of published materials are valuable as additional reference materials, including both books and periodicals.

Books

The following books may be worthwhile sources of additional information:

Ferret, by F. P. Miller, A. F. Vandome and J. M. Brewster, 2009. Alphascript Publishing.

The Ferret: An Owner's Guide to a Happy, Healthy Pet, by M. R. Shefferman, 2001. Wiley Publishing.

Formulary for Laboratory Animals, 3rd edition, by C. T. Hawk, S. Leary, and T. Morris, 2005. Wiley-Blackwell, Inc.

Clinical Laboratory Animal Medicine: An Introduction, 3rd edition, by K. Hrapkiewicz and L. Medina, 2006. Wiley-Blackwell, Inc.

The IACUC Handbook, 2nd edition, edited by J. Silverman, M. A. Suckow, and S. Murthy, 2007. CRC Press/Taylor & Francis Group.

Making and Using Antibodies: A Practical Handbook, edited by G. C. Howard and M. R. Kaser, 2007. CRC Press/Taylor & Francis Group.

Anesthesia and Analgesia in Laboratory Animals, 2nd edition, edited by R. Fish, P. Danneman, M. Brown, and A. Karas, 2008. Academic Press, Elsevier, Inc.

Periodicals

Comparative Medicine. Published by AALAS.

Journal of the American Association for Laboratory Animal Science. Published by AALAS.

Laboratory Animals. Published by Royal Society of Medicine Press (www.rsmjournals.com).

Lab Animal. Published by Nature Publishing Group (www.labanimal.com).

ILAR Journal. Published by the Institute for Laboratory Animal Research (www.dels.nas.edu/ilar).

ALN Magazine. Published by Vicon Publishing, Inc. (www.animallab.com).

electronic resources

CompMed. Available through AALAS (www.aalas.org), CompMed is a listserve for biomedical research. CompMed is limited to participants who are involved in some aspect of biomedical research or veterinary medicine, including veterinarians, technicians, animal facility managers, researchers, and graduate/veterinary students. AALAS membership is not required to subscribe to CompMed.

TechLink. Also available through AALAS, TechLink is a listserve created especially for animal care technicians in the field of laboratory animal science. Open to any AALAS national member, TechLink serves as a way for laboratory animal technicians to exchange information and conduct discussions of common interest via e-mail messages with technicians in the United States and other countries around the world.

IACUC.ORG. Produced by AALAS, IACUC.ORG (www.IACUC.org) is an information resource for members and staff of institutional animal care and use committees. It is a link archive where online resources are organized by menus and submenus. IACUC.ORG was developed as an organizational tool to quickly point to a topic of interest, such as protocol forms or disaster plans used by various institutions.

AALAS Learning Library. The AALAS Learning Library provides training that is essential for technicians, veterinarians, managers, IACUC members, and investigators working with animals in a research or education setting. The courses emphasize the appropriate handling, care, and use of animals, including ferrets.

LAWTE Listserve. The Laboratory Animal Welfare Training Exchange (www.lawte.org) maintains a members-only listserve where individuals may seek information from colleagues. In addition, the site maintains a library of materials relevant to training of individuals in the proper handling and care of laboratory animals.

references

1. HMS Ferret. Various Authors. 15 Feb 2011. http://en.wikipedia.org/wiki/HMS_Ferret. Retrieved: 31 May 2011.

2. Matsuoka Y., E.W. Lamirande and K. Subbarao. 2009. The ferret model for influenza. *Current Protocols in Microbiology* May: Chapter 15: Unit 15G.2.

3. Zitzow L.A., T. Rowe and T. Morken, et al. 2002. Pathogenesis of avian influenza A (H5N1) viruses in ferrets. *J Virol* 76: 4420–29.

4. Li Z. et al. 2006. Cloned ferrets produced by somatic cell nuclear transfer. *Dev Biol* 293(2): 439–48.

5. Sun X, H. Sui and J.T. Fisher, et al. 2010. Disease phenotype of a ferret CFTR-knockout model of cystic fibrosis. *J Clin Invest* 120. 3149–60.

6. Johnson-Delaney, C.A. 2005. The ferret gastrointestinal tract and *Helicobacter mustelae* infection. *Vet Clin N Am–Exot Anim Prac* 8: 197–212.

7. Fox, J.G. 1998. *Biology and Diseases of the Ferret*. 2nd ed. Baltimore: Williams and Wilkins.

8. Manning D.D. and J.A. Bell. 1990. Lack of detectable blood groups in domestic ferrets: Implications for transfusion. *J Am Vet Med Assoc* 197(1): 84–86.

9. Marshall Bioscience. 2011. URL: http://www.marshallbio.com/

10. Johnson-Delaney, C.A. 1996. *Ferrets: An Exotic Animal Companion Handbook for Veterinarians*. Ed Harrison. Lake Worth: Wingers Publishing.

11. Platt, S.R. et al. 2004. Composition of cerebrospinal fluid in clinically normal adult ferrets. *Am J Vet Res* 65(6): 758–60.

12. United States Department of Agriculture. 2005. *Animal Welfare Act and Animal Welfare Regulations*. U. S. Government Printing Office.

13. Gad, S. (Ed). 2007. The ferret. In *Animal Models in Toxicology*. Boca Raton: Taylor & Francis.

14. Queensbury, K.E. and J.W. Carpenter. 2004. *Ferrets, Rabbits, and Rodents Clinical Medicine and Surgery*. 9–11. St. Louis: Saunders.

15. American Association for Laboratory Animal Science. 2009. Less common species used in research. *Assistant Laboratory Animal Technician Training Manual*. 209–11. USA: Drumwright & Co.

16. Morton, E.L. and C. Mathis. 2010. *Ferrets: A Complete Owner's Manual*. Hauppauge: Barron's Educational Series, Inc.

17. Shilling, K. 2007. *Ferrets for Dummies*. Indianapolis: Wiley Publishing, Inc.

18. National Research Council of the National Academies. 2011. *Guide for the Care and Use of Laboratory Animals*, 8th ed. Washington, DC: National Academies Press.

19. Scipioni Ball, R. 2006. Issues to consider for preparing ferrets as research subjects in the laboratory. *ILAR Journal* 47:4 348–57.

20. Suckow, M.A. and V. Schroeder. 2010. Husbandry. In *The Laboratory Rabbit*. 11–31. Boca Raton: CRC Press.

21. Public Health Service. 2002. Public Health Service Policy on Humane Care and Use of Laboratory Animals. Publication of the Department of Health and Human Services, National Institutes of Health, Office of Laboratory Animal Welfare. Available at: http://grants.nih.gov/grants/olaw/references/phspol.htm; accessed April 4, 2011.

22. Health Research Extension Act of 1985. Public Law 99–158, Section 495.

23. FDA 21CFR58. 1978 and all Amendments. Available at: http://www.accessdata.fda.gov/scripts/cdrh/cfdocs/cfcfr/CFRSearch.cfm?CFRPart=58; accessed April 5, 2011.

24. Kusters, I.C., J. Matthews and J.F. Saluzzo. 2009. Manufacturing vaccines for an emerging viral infection. *Vaccines for Biodefense and Emerging and Neglected Diseases*. Burlington, MA: Academic Press.

25. EPA 40CFR792. 1989 and all Amendments. Available at: http://ecfr.gpoaccess.gov/cgi/t/text/text-idx?c=ecfr&tpl=/ecfrbrowse/Title40/40cfr792_main_02.tpl; accessed April 5, 2011.

26.Hornshaw, T.C, R.K. Ringer and R.J. Aulerich, et al. 1986. Toxicity of sodium monofluroacetate (Compound 1080) to mink and European ferrets. *Environ Toxicol Chem* 5(2): 213–23.

27.Hornshaw, T.C., R.J. Aulerich and R.K. Ringer. 1986. Toxicity of o-cresol to mink and European ferrets. *Environ Toxicol Chem* 5(8): 713–20.

28.Federation of Animal Science Societies. 2010. *Guide for the Care and Use of Agricultural Animals in Research and Teaching*, 3rd edition. Champaign, IL: FASS.

29.Council of Europe. 1986. European Convention for the Protection of Vertebrate Animals Used for Experimental and Other Scientific Purposes.

30.Ball, R.S. 2006. Issues to consider for preparing ferrets as research subjects in the laboratory. *ILAR Journal* 47(4): 348–57.

31.California Fish and Game Code, Sections 2116–2126.

32.Hawaii Department of Agriculture. Animal Guidelines for Importation to Hawaii. Available at: http://hawaii.gov/hdoa/ai/pi/pq/animal; accessed April 5, 2011.

33.Morrison, G. 2001. Zoonotic infections from pets: Understanding the risks and treatment. *Postgrad Med* 110(1): 29–30, 35–36.

34.Mitchell, M.A. and T.N. Tully. 2004. Chapter 40, Zoonotic diseases. *Ferrets, Rabbits and Rodents: Clinical Medicine and Surgery*, 2nd ed. St. Louis: Saunders.

35.Fox, J.G., J.L. Ackerman and N.S. Taylor. 1987. *Campylobacter jejuni* in the ferret: A model of human campylobacteriosis. *Am J Vet Res* 48: 85–90.

36.Marini, R.P., J.A. Adkins and J.G. Fox. 1989. Proven or potential zoonotic diseases of ferrets. *J Am Vet Med Assoc* 195: 990–94.

37.Rehg, J.E., F. Gigliotti and D.C. Stokes. 1988. Cryptosporidiosis in ferrets. *Lab Anim Sci* 38: 155–158.

38.Abe, N., C. Read and R.C. Thompson, et al. 2005. Zoonotic genotype of *Giardia intestinalis* detected in a ferret. *J Parasitol* 91(1): 198–202.

39.Hagen, K.W. and J.R. Gorham. 1972. Dermatomycoses in fur animals: Chinchilla, ferret, mink and rabbit. *Vet Med Small Anim Clin* 67(1): 43–48.

40.More, D. About.com. Allergies. January 31, 2009. Available at: http://allergies.about.com/od/animalandpetallergy. Retrieved: May 31, 2011.

41.Allergic Reaction to Ferrets: Treatment and Avoidance. Various Authors. May 2, 2011. Available at: http://allergy.reliefsource.com/pet-allergies. Retrieved: May 31, 2011.

42.Lloyd, M. 2002. Veterinary care of ferrets 1. Clinical examination and routine procedures. *In Practice* 24: 90–95.

43.McLain, D.E. 2006. *Toxicology: Animal Models in Toxicology*. Boca Raton: CRC Press.

44.Rylan, L.M. and E. Lipinsky. 1994. A technique for vasectomizing male ferrets. *Canine Practice* 19: 135–26.

45.Schoemaker, N.J. 2010. What every veterinarian should know about ferret medicine. *Proceedings of the European Veterinary Conference*. Voorjaarsdagen. Amsterdam, Netherlands.

46.Quesenberry, K.E. 1997. *Ferrets, Rabbits and Rodents*. Philadelphia: Saunders Company.

47.Lloyd, M. 1999. *Ferrets: Health, Husbandry and Diseases*. Wiley-Blackwell.

48.Williams, B.H. 2000. Therapeutics in ferrets. *Vet Clin N Am–Exot Anim Prac* 3 N1:131–53. Saunders Company.

49.Wagner, R.A. 2009. Ferret cardiology. *Vet Clin North Am–Exot Anim Prac* 12:115–34.

50.Besch-Williford, C.L. 1987. Biology and medicine of the ferret. *Vet Clin N Am–Small Anim Prac (Exotic pet medicine)* 17(5): 1155–83.

51.Diaz-Figueroa, O. and M.O. Smith. 2007. Clinical neurology of ferrets. *Vet Clin North Am–Exot Anim Prac* 10: 759–73.

52.Ivey, E. and J. Morrisey. 1999. Ferrets: Examination and preventive medicine. *Vet Clin N Am–Exot Anim Prac* 2: 471–94.

53.Oglesbee, B.L. 2006. *The Five-Minute Veterinary Consult: Ferret and Rabbit*. Blackwell Publishing.

54.Lennox, A.M. 2007. How I manage diarrhea and weight loss in ferrets. NAVC Conference Proceedings (North American Veterinary Conference). Reprinted in the IVIS website with permission of the SEVC. www.ivis.org.

55.Kirk, R.W. and J.D. Bonagura, ed. 1992. *Kirk's Current Veterinary Therapy XI: Small Animal Practice*. Philadelphia: Saunders Company.

56.Hoppes, S.M. 2010. The senior ferret (*Mustela putorious furo*). *Vet Clin N Am–Exot Anim Prac* 13: 107-22.

57.Schoemaker, N.J. 2010. Approach to the ferret with hind-limb weakness. *Proceedings of the European Veterinary Conference-*Voorjaarsdagen-Amsterdam, Netherlands.

58.Otto, G., D.W. Rosenblad and J.G. Fox. 1993. Practical venipuncture techniques for the ferret. *Laboratory Animals*. 27: 26–29.

59.Fish, R. and P.J. Dannenman (Eds.). 2008. *Anesthesia and Analgesia in Laboratory Animals*, 2nd ed. San Diego: Academic Press.

60.Sylvina, T.J., N.G. Berman and J.G. Fox. 1990. Effect of yohimbine on bradycardia and duration of recumbency in ketamine-xylazine anesthetized ferrets. *Lab Anim Sci* 40: 178–82.

61.Schwarz, L.A., M. Solano and A. Manning, et al. 2003. The normal upper gastrointestinal examination in the ferret. *Vet Radiol and Ultrasound* 44: 165–72.

62.Flecknell, P. 2009. *Flecknell's Laboratory Animal Anesthesia*, 3rd ed. Academic Press.

63.Kircher, S.S., L.E. Murray and M.L. Juliano. 2009. Minimizing trauma to the upper airway: A ferret model of neonatal intubation. *JAALAS* 48: 780–84.

64.Evans, T. and K.K. Springsteen. 1998. Anesthesia of ferrets. *Seminars in Avian and Exotic Pet Medicine* 1: 48–52.

65.Mitchell, M.A. and T.M. Tully. 2008. *Manual of Exotic Pet Practice*. Elsevier.

66.Abou-Madi, N. 2006. *Recent Advances in Veterinary Anesthesia and Analgesia: Companion Animals*. Ithaca: International Veterinary Information Service (IVIS). www.ivis.org

67.ACVA Monitoring Guidelines Update. 2009. Recommendations for monitoring anesthetized veterinary patients.

68.Busch, S.J. 2006. *Small Animal Surgical Nursing: Skills and Concepts*. Elsevier Mosby.

69.Bennett, A. 2009. Soft surgery in ferrets. *Proceedings of the SEVC (Southern European Veterinary Conference)*. Reprinted in the IVIS website with permission of the SEVC. www.ivis.org

70.Marini, R.P. and J.G. Fox. 1998. *Biology and Diseases of the Ferret*, 2nd ed. Lippincott Williams & Wilkins.

71. American Veterinary Medical Association. 2007. *AVMA Guidelines on Euthanasia* (formerly: *Report of the AVMA Panel on Euthanasia*).

72. Fitzhugh, D.C., A. Parmer and L.I. Shelton, et al. 2008. A comparative analysis of carbon dioxide displacement rates for euthanasia in ferrets. *Lab Anim* 37: 81–86.

73. Kahn, C.M. and S. Line. 2006. *The Merck Veterinary Manual*, 9th ed. Whitehouse Station: Merck & Co.

74. Torres-Medina, A. 1987. Isolation of an atypical rotavirus causing diarrhea in neonatal ferrets. *Lab Anim Sci* 37: 167–71.

75. Lightfoot, T.L. 2009. Gastrointestinal presentations in ferrets (How I treat Inflammatory Bowel Disease). *Proceedings of the SEVC (Southern European Veterinary Conference)*. Reprinted in the IVIS website with permission of the SEVC. www.ivis.org

76. Fox, J.G. and R.P. Marini. 2001. *Helicobacter mustelae* infection in ferrets: Pathogenesis, epizootiology, diagnosis and treatment. *Seminars in Avian and Exotic Pet Medicine* 10: 36–44.

77. Williams, B.H., M. Kiupel and K.H. West, et al. 2000. Coronavirus: Associated epizootic catarrhal enteritis in ferrets. *J Am Vet Med Assoc* 15: 526–30.

78. Lloyd, M. 2002. Veterinary care of ferrets 2: Common clinical conditions. *In Practice* 24: 136–45.

79. Chandra, S. 2006. Pathology. In *Animal Models in Toxicology*. Boca Raton: CRC Press.

80. Laprie, C., J. Duboy and J. Martinez. 2009. Systemic coronavirus-associated disease in the domestic ferret (*Mustela putorious*): Hystopathologic and immunohistochemical characterization in three ferrets. *Pratique Medicale et Chirurgical de l'Animal de Companie* 44: 111–15.

81. Murray, J., M. Kiupel and R.K. Maes. 2010. Ferret coronavirus: Associated diseases. *Vet Clin N Am–Exot Anim Prac: Advances and Updates in Internal Medicine* 13: 543–60.

82. Charles, S.D., H.M. Chopade and D.K. Ciszewski, et al. 2007. Safety of 5% ponazuril (toltrazuril sulfone) and efficacy against naturally acquired *Cystoisospora ohiensis*-like infection in beagle puppies. *Parasitol Res* 101: S137–S144.

83. Mitchel, M.A. 2008. Ponazuril. *J Exot Pet Med* 17: 228–29.

84. Reinmeyer, C.R., D.S. Lindsay and S.M. Mitchell, et al. 2007. Development of experimental *Cystoisospora canis* infection models in beagle puppies and efficacy evaluation of 5% ponazuril (toltrazuril sulfone) oral suspension. *Parasitologic Res* 101:129–36.

85. Marini, R.P., J.G. Fox and N.S. Taylor, et al. 1999. Ranitidine bismuth citrate and clarithromycin alone or in combination, for eradication of *Helicobacter mustelae* infection in ferrets. *Am J Vet Res* 60: 1280–85.

86. Johnston, M.S. 2005. Clinical approaches to analgesia in ferrets and rabbits. *Seminars in Avian and Exotic Pet Medicine* 14: 229–35.

87. Schaffer, D. 1994. Miscellaneous species: Anesthesia and analgesia. *Research Animal Anesthesia, Analgesia and Surgery*. Greenbelt: Scientists Center for Animal Welfare.

88. Murat, I., R.J. Callahan and L.R. Jackson. 1988. Minimum alveolar concentrations (MAC) of halothane, enflurane and isoflurane in ferrets. *Anesthesiology* 68: 783–86.

89. Lawson, A. et al. 2006. Comparison of sevoflurane and isoflurane in domestic ferrets. *Vet Therapeutics* 7(3): 207–12.

90. Marini, R.P., L.R. Jackson and M.L. Esteves, et al. 1994. Effect of isoflurane on hematologic variables in ferrets. *Am J Vet Res* 55: 1479–83.

91. Mac Phail, C.M., E. Monnet and J.S. Gaynor, et al. 2004. Effect of sevoflurane on hemodynamic and cardiac energetic parameters in ferrets. *Am J Vet Res* 65: 635–58.

92. Marini, R.P. et al. 1994. Effect of isoflurane on hematologic variables in ferrets. *Am J Vet Res* 55(10): 1479–83.

93. Marini, R.P. et al. 1997. Distribution of technetium 99m-labeled red blood cells during isoflurane anesthesia in ferrets. *Am J Vet Res* 58(7): 781–85.

94. Florczyk, A.P. and J.E. Schurig. 1981. A technique for chronic jugular catheterization in the ferret. *Pharmacol Biochem Behav* 14(2): 255-57.

95. Greener, Y. and Gilles, B. 1985. Intravenous infusion in ferrets. *Lab Anim* 14: 41–44.

96. Mesina, J.E. et al. 1988. A simple technique for chronic jugular catheterization in ferrets. *Lab Anim Sci* 38(1): 89–90.

97. Fox, J.G. 2002. *Laboratory Animal Medicine.* 2nd ed. American College of Laboratory Animal Medicine. Amsterdam, New York: Academic Press.

98. Meredith, A. and S. Redrobe. 2002. *Manual of Exotic Pets.* 4th ed. BSAVA Manuals Series. Quedgeley: BSAVA.

99. Lee, E.J. et al. 1982. Haematological and serum chemistry profiles of ferrets (*Mustela putorius furo*). *Lab Anim* 16(2): 133–37.

100. Brown, C. 2006. Blood collection from the cranial vena cava of the ferret. *Lab Anim* 35(9): 23–24.

101. Otto, G., W.D. Rosenblad and J.G. Fox. 1993. Practical venipuncture techniques for the ferret. *Lab Anim* 27(1): 26–29.

102. Morrisey, J.K. and J.C. Ramer. 1999. Ferrets. Clinical pathology and sample collection. *Vet Clin N Am–Exot Anim Prac* 2(3): 553–64.

103. Joslin, J.O. 2009. Blood collection techniques in exotic small mammals. *J Exot Pet Med* 18: 117–39.

104. Marini, R.P., M.I. Esteves and J.G. Fox. 1994. A technique for catheterization of the urinary bladder in the ferret. *Lab Anim* 28(2): 155–57.

105. Platt, S.R. et al. 2004. Composition of cerebrospinal fluid in clinically normal adult ferrets. *Am J Vet Res* 65(6): 758–60.

106. Poole, T.B., R. Robinson, and Universities Federation for Animal Welfare. 1986. *The UFAW Handbook on the Care and Management of Laboratory Animals.* 6th ed. London, New York: Longman, Churchill Livingstone.

Index

A

U

V

W

Y

Z

notes

notes

notes

notes

notes

9781439861813